AGROFORESTRY IN DRYLAND AFRICA

AGROFORESTRY

Illustrated by TERRY HIRST

International Council for Research in Agroforestry

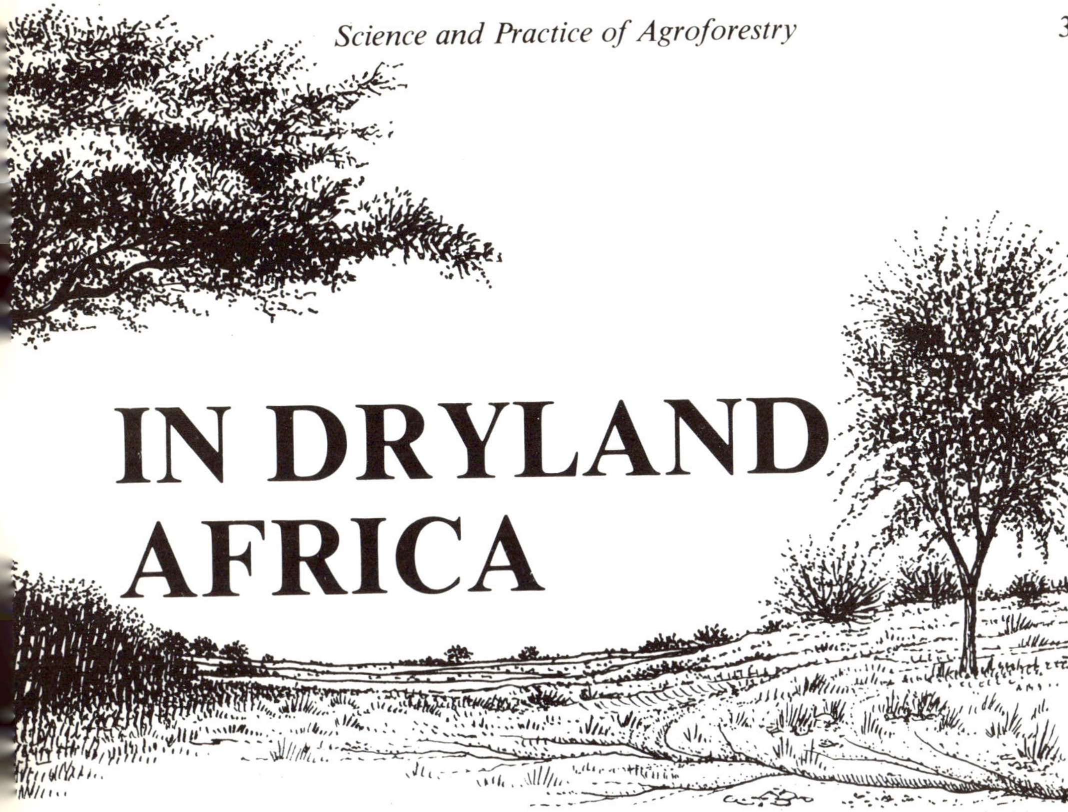

IN DRYLAND AFRICA

Written by DIANNE ROCHELEAU
FRED WEBER
ALISON FIELD-JUMA

ICRAF

NAIROBI 1988

Published by the International Council for Research in Agroforestry
ICRAF House, off the Limuru Road, P. O. Box 30677, Nairobi, Kenya

Printed by English Press
P. O. Box 30127, Nairobi, Kenya

Designed and illustrated by Hirst Studio Productions
P. O. Box 14432, Nairobi, Kenya

ISBN 92 9059 049 1

The preparation and production of this book were supported through a generous grant from the Swedish International Development Authority (SIDA).

FOREWORD

This book, *Agroforestry in Dryland Africa*, is the third in a series published by ICRAF on the science and practice of agroforestry. The series is intended to include practical handbooks and manuals, descriptions of research methods, monographs on multipurpose tree species, analyses of specific agroforestry practices and systems, and reviews of special aspects of agroforestry.

Agroforestry in Dryland Africa is specifically designed to be used by agroforestry field workers in subhumid and semi-arid regions of the continent. This includes researchers working with grassroots organizations, community extension and development workers and development specialists in government agencies and national research institutes. It will also be useful to extension training officers and people engaged in liaison between extension and research on the use of natural resources. Its subject matter assures it of a wide readership in other parts of the world.

Agroforestry—comprising all land-use systems and practices in which woody perennials are deliberately grown on the same land-management unit as crops and/or animals—is a rapidly evolving approach to resource management. It involves international research centres, large-scale farming and forestry enterprises, government extension services, community development agencies and local farmers and livestock owners. Although agroforestry is now widely practiced and studied throughout the world, there are few written materials which directly link research and practice. Thus, this book fills a gap in the information available for use in the field.

The material in *Agroforestry in Dryland Africa* is drawn from the accumulated experience of practitioners and researchers. It provides an introduction to agroforestry and the roles of various agroforestry practices in the community setting. This is followed by a description of the process of working with community members to identify and adapt agroforestry practices to meet local needs. Fifteen agroforestry practices are described in Part II with examples from dry regions of Africa. Part III consists of tools for the practitioner: information on trees and shrubs suited to the region and guidelines for assessing community needs and designing appropriate agroforestry activities. A list of contacts in the region is also provided to facilitate and encourage communication of research results and experience in the field.

The preparation and production of this book were supported through a generous grant from the Swedish International Development Authority (SIDA).

B. Lundgren
Director-General, ICRAF

PREFACE

Agroforestry in Dryland Africa presents one approach to developing agroforestry practices suitable for the subhumid and semi-arid regions of Africa. It provides a framework for working with community members to evaluate current land-use systems and to develop sustainable improvements using agroforestry techniques. We have attempted to provide the practical information needed in order to implement soil and water conservation measures which will also satisfy the numerous other needs of rural communities. We describe the management and multiple uses of tree species suited to the region and discuss the design, management and potential benefits of 15 agroforestry practices. This discussion includes issues of land and tree tenure, the distribution of benefits and participation of all community members.

The most common approach to technical innovation is still for researchers to design technologies and then allow 'progressive' members of the community to adopt them and make minor adjustments. We suggest trying the reverse: a variety of local land users (not just well-to-do farmers) develop ideas for innovation together with community-based research and development workers. In this way, the technologies developed are more likely to reflect the diversity of local priorities and needs. This approach also promotes the accountability of researchers and development workers to the local community of land users.

The technologies developed in one area can be extended and adapted by other people in similar sites. They may also be tested and refined by formal research institutions for use under a broader range of conditions.

Through this process, researchers and development workers can add their own special expertise to that of local farmers and other land users and can function as two-way information channels between rural land users, national and international development agencies and the scientific research community. They can document a wide variety of experience with both the technical and social aspects of agroforestry systems which will help to ensure that future research truly responds to people's needs. In this way, agroforestry researchers and development workers can make a valuable contribution to the formulation and implementation of broader programmes aimed at ensuring the sustainable use of natural resources.

It is the authors' hope that this book, which covers over half of the African continent, will be adapted to individual countries and regions and translated into local languages. The continent is too vast and varied for one volume to describe its ecological, cultural and economic systems in sufficient detail. As a start, Part III of this book has been designed to be adapted by the user to include local information.

ACKNOWLEDGEMENTS

We are grateful to a large number of friends and colleagues who have assisted us with this book from planning through to production. We are indebted to several reviewers who made extensive comments or contributed material for use in the text: Ed Barrow, Marilyn Hoskins, Paul Kerkoff, the field team of CARE Kenya Agroforestry Projects, Oscar Oyalo, Peter Wood, Michael Baumer, Denis Depommier, Peter von Carlowitz and Tony Young. John Raintree, Sara Scherr, Calestous Juma, Louise Buck and Luis Malaret provided ideas, moral support and valuable critical comments.

Terry Hirst and Nereas Giconi contributed their time, ideas and creativity far beyond their role in design and illustration. Richard Labelle brought Fred Weber's original proposal to ICRAF and contributed extensively to the search for information, while Sidney Westley edited the text, brought together the different sections and shepherded the volume through final production. Stephen Okemo and Gregor Wolf also helped with literature and data-base searchers. Special thanks go to Stella Muasya for her substantial efforts in typing and re-typing our cryptic drafts; her patience and commitment to this work greatly assisted in its production.

Our best critics and sources of inspiration have come from the field—in remote areas throughout Africa, from the Sahel to the highlands of Rwanda. Closer to home, we have drawn heavily on the experience and judgement of the staff and participants of ICRAF's Kathama Project: Japheth Kyengo, Veronica Ndunge, Lawrence Kyongo, Remko Vonk, Bwana Musyoki and several women's self-help groups. And finally, thanks to Ramon Rocheleau Malaret and the Ford Foundation for allowing Dianne Rocheleau time to complete work on this manuscript.

Dianne Rocheleau, Fred Weber
and Alison Field-Juma
Nairobi, March 1988

CONTENTS

PART ONE

AGROFORESTRY AS A POPULAR SCIENCE

The science of agroforestry still lies mainly in field practice and in systems of knowledge outside the formal scientific community. Professional scientists have contributed to the knowledge and improvement of particular agroforestry species and techniques. However, there is a much broader scope of agroforestry knowledge and practice in the field and a great need for work to bridge the gap between popular science and practice and the more formal results of institutionalized scientific research. The three chapters in this section address the diversity of existing agroforestry practices and their potential contribution to sustainable development in dryland Africa, as well as the potential for developing and improving agroforestry in partnership with rural people.

Chapter 1 presents an overview of agroforestry, a brief description of 15 specific practices and an introduction to a community-based approach to agroforestry research and extension. Chapter 2 offers a rationale and guidelines for a series of activities to gather, discuss and act on agroforestry information in partnership with participating communities. Finally, Chapter 3 contains suggestions for monitoring and evaluating both the practices and the process involved in agroforestry development. This section should provide the understanding and motivation to make the best use of the more specific information contained in Parts II and III.

CHAPTER ONE

INTRODUCTION TO AGROFORESTRY

1.1 What is Agroforestry?

Agroforestry is a popular concept among agricultural development and environmental specialists and is often invoked by scientists and planners as a solution to rural development needs in Africa. While the word and the optimism associated with it are widely shared, the actual meaning of agroforestry is often misunderstood. Agroforestry, as used in this manual, is defined as all practices that involve a close association of trees or shrubs with crops, animals and/or pasture. This association is both ecological and economic. Agroforestry may involve a combination of practices in the same place at the same time (intercropping and related practices), or practices in the same place but at different times (rotational practices). The 'place' may be as small as a single garden or

Date palms over sorghum, and a typical home garden.

Boran bull under acacia shade at midday.

cropland plot or as extensive as a small watershed or a vast stretch of communal grazing land.

Some identify agroforestry with a particular practice or set of practices, for example growing crops between hedgerows to improve soil fertility or growing a variety of species in multistorey home gardens. Agroforestry is also often associated with the cropping systems and environments of the humid tropics. In some cases, this has led to agroforestry projects based on species and practices that do not meet the needs and conditions of people living in other types of environment.

If agroforestry is to serve people's needs in a variety of rural settings, it is important to see it as an approach to land use, rather than as a fixed arrangement of plants or a particular combination of species. A brief review of the full range of agroforestry practices will provide a basis for thinking about agroforestry systems that could usefully be introduced in different environments, including the dryland areas of Africa. These can be grouped into practices involving trees with crops, trees with pastures, trees with animals and trees nested into special places in the landscape. These agroforestry practices are described in greater detail in Part II of this book.

AGROFORESTRY PRACTICES IN CROPLAND

One of the most widely practiced agroforestry systems in the dryland areas of Africa is based on trees dispersed in cropland (section 4.1). The trees are usually permanent and full sized and they may be dispersed either singly or in clumps. In some cases, farmers plant or maintain trees in their cropland primarily to obtain valuable tree products. In other cases, the trees seem to increase the production of the surrounding crops and improve the soil and water conditions for crop growth.

Contour vegetation strips with multipurpose trees and tree crops (section 4.2) provide another example of an agroforestry practice that fits well into current farming systems in much of subsaharan Africa. Contour vegetation strips are usually introduced in order to prevent soil erosion on sloping croplands, while at the same time providing useful products such as food, fodder or wood. These living barriers may consist of grasses or ground cover only, but it is often desirable to include trees and shrubs.

Multipurpose trees, grasses and other herbaceous plants are often combined along the edges and uncultivated spaces of soil and water conservation structures, ranging from small contour bunds and ditches (section 5.1) to bench terraces (section 5.2) on cropland. These plant combinations can produce useful items for home use or sale, while helping to stabilize and protect conservation structures from direct exposure to rain and wind.

Alley cropping, or hedgerow intercropping, is perhaps the best known but least understood of all agroforestry practices used on cropland. While there are many variations, alley cropping most often consists of dense hedges of multipurpose trees planted in rows between wider strips of annual crops. The hedges are lopped to produce mulch, which is applied to the cropped areas to fertilize and cover the soil.

Another arrangement is multistorey, closely spaced trees intercropped with annual plants. In contrast to dispersed trees in cropland, this arrangement is often based on shade-tolerant understorey crops and on a greater diversity of tree and hedgerow species. It resembles home gardens, except that it usually occurs in cropland and the trees are more widely spaced. While this practice is more common in humid areas, it may occur in the drier zones of Africa in both rainfed and irrigated croplands.

The practice of mulching, composting or mounding cropland with tree leaves does not necessarily require the presence of trees in cropland, but it is still an agroforestry practice, using tree leaves to protect and improve the soil and to increase crop yields. In the case of mulching, leaves are applied directly to the soil. In composting systems, leaves and twigs are combined with manure and processed into an organic fertilizer which is applied later. In mounding systems, the tree leaves and twigs are combined with grasses in long mounds in cropland.

The mounds are covered with a layer of soil and left to decompose in place and crops are planted into the mounds in the next season.

AGROFORESTRY PRACTICES IN FALLOW CROPLAND

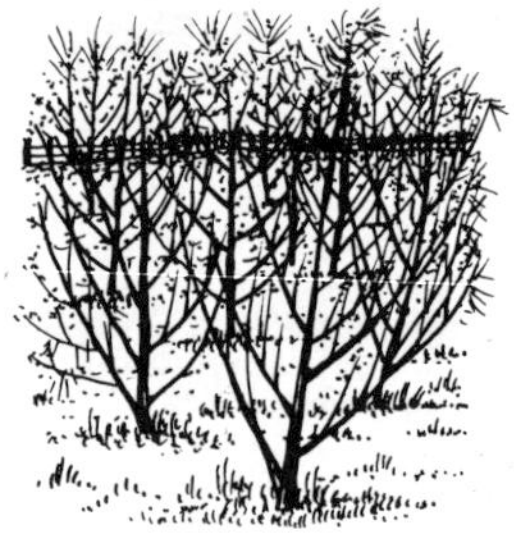

Fallows are croplands left without crops for periods ranging from one season to several years. The objective is to control insect pests, diseases and weeds associated with previous cropping and to recover depleted soil nutrients. Once the soil has recovered, crops are reintroduced for one or more seasons, after which the fallow is repeated. Improved fallows (section 4.5) may involve only the selective cutting and weeding of the natural vegetation, additions to the natural vegetation or even the replacement of the natural vegetation with trees, herbaceous plants or animals. Normally, improved fallows can be expected to restore the cropland more quickly than natural fallows, allowing a shorter fallow period before cropping begins again. It is also possible to introduce permanent trees and shrubs which will be maintained through future cycles of cropping and fallows.

AGROFORESTRY PRACTICES IN PASTURES AND RANGELAND

Sylvopastoral systems (Chapter 7), combining woody plants with grasses and other herbaceous fodder plants, are widespread throughout subsaharan Africa. Extensive sylvopastoral systems on rangeland usually involve the selective protection and management of naturally occurring trees and shrubs of particular value for animal fodder. Trees may also be purposely planted with existing grasses, either dispersed as individuals, in clumps or in rare cases in lines or blocks. In addition to high-protein fodder for livestock, the trees may provide building poles, fuelwood, fruit or cash crops such as resins.

More intensive sylvopastoral systems are found in natural or improved pastures in farming areas. Naturally occurring trees may be managed selectively or multipurpose trees and fodder shrubs may be planted, either dispersed as individuals, in lines, along contours or in clumps or blocks. Tree products range from timber to fruit, fuelwood and high-protein fodder. In these situations, trees are spaced more closely and managed more intensively than in extensive systems found in rangelands.

AGROFORESTRY PRACTICES ON BOUNDARIES AND BORDER SPACES

Living fences and living fenceposts (section 6.1) are used throughout Africa to protect people and their dwellings, crops, animals and other

property. They may be designed to fence animals in or to keep people and animals out of a particular space. Plants may form the entire fence structure, or living trees may be used as fenceposts while the rest of the fence is made of wire or dead branches and reeds.

Boundary markers (6.2) are different from living fences, as their main purpose is to make boundaries clear, without necessarily enforcing them. Even where it is not important to mark boundaries, boundary spaces may provide a convenient site for planting productive trees and shrubs that do not fit in with other land uses elsewhere. A few large trees of a particular kind may be planted as individuals, or timber or multipurpose trees may be planted in lines or in dense hedges.

Windbreaks (6.3) are often, but not always, located on boundaries between properties or fields. Their main function is to protect homes, crops, pastures and soil and water resources from damage by wind. They may take many forms—from large shelterbelts for whole villages to individual windbreak strips for one field or a single homestead. Windbreaks are normally multistoried, including one or more rows of trees placed across the path of the prevailing winds.

AGROFORESTRY PRACTICES ALONG WATERWAYS

Floodplain gardens (section 6.4) are located in isolated depressions, along the flatter and more stable portions of river and stream banks or on the edges of lakes and ponds. These gardens often include trees, shrubs and woody vines as well as vegetable crops, medicinal plants, spices and root crops. Such plots are often located on public or communally owned land where well managed private or group gardens are tolerated. These sites have a unique production potential because of their access to water and fertile soils.

Multipurpose trees, shrubs and grasses may be planted to help stabilize rock and wooden structures for erosion control across gully channels (section 5.1). They may also be planted in lines to form living structures across the lower reaches of shallow channels or to help to stabilize the areas behind erosion-control structures once these areas have filled in with soil and debris. Once stabilized, such sites may be highly productive because of the controlled drainage of surface and subsurface water into the filled sections behind the structures. Timber, fuelwood and fodder can all be produced from woody plants growing in these sites and in some cases farmers may eventually develop small fruit and vegetable plots.

Multipurpose trees and tree crops may also be established with grasses on the sloping banks of streams, gullies or channels (sections 5.3

and 6.4). In such sites, they serve to protect the soil on the slopes, to shade the watercourses and to provide fuelwood, fodder, fruit or other products which do not require removal of the trees or ground cover.

AGROFORESTRY PRACTICES IN HOME COMPOUNDS

Home gardens (section 4.4) occur in some form in almost every ecological zone and farming system in Africa. Agroforestry practices in home gardens can range from a few trees and shrubs in a small vegetable and herb garden to a dense multistoreyed plot of fruits, vegetables, herbs and cash crops with trees planted for timber, fuelwood and/or fodder. A home garden may serve as a specialized plot within a larger production system or it may represent the main cultivated plot and a major source of food and cash income, especially for a poor family with little arable land.

Decorative and shelter planting (section 6.6) around houses may also include agroforestry practices. For example, fruit-bearing vines may be mixed with large ornamental trees or vegetable gardens may be combined with rows of flowering or decorative trees. The bases of large shade trees or fenced fruit trees can provide safe, convenient spaces for small nurseries of tree and vegetable seedlings. Living fences can also be an integral part of a home compound, whether to control animals or simply to define spaces for different purposes. These are often decorated with flowering or fruit-bearing vines or shrubs. Even when home-compound plantings do not directly include agroforestry practices, they can provide a testing ground and display case for new agroforestry species and techniques for tree establishment and management.

AGROFORESTRY PRACTICES IN PUBLIC AND SHARED SPACES

Decorative, symbolic and shelter planting in public places (see section 6.6) may take many forms, most of which do not include agroforestry practices. However, such places may be excellent sites to demonstrate new agroforestry practices or species to the local community. The use of woody plants in public spaces may range from a single large tree of religious or cultural significance to a public flower garden, which could include one or more ornamental trees. Trees that provide shade, fruit or fodder may be planted in sites such as public markets, wells, clinics or places of worship.

Public spaces also include community plots for the production of wood, fodder, food or cash crops. These may take the form of woodlots, plantations or gardens and may combine trees and shrubs with animals or herbaceous plants. Most community plots have the potential to include agroforestry practices, either within the plot or as living fences.

Roadside plantings (6.5) resemble other plantings in public places. They may include ornamental and shade trees or trees that provide useful products for local consumption. Roadside plantings may also be used to demonstrate agroforestry species and practices. Roadsides are particularly well suited to combinations of grasses and trees or fully developed agroforestry production systems. Government agencies or residents may plant trees along roadsides for shade, fodder, fuel, oilseeds, fruit or other products. In many situations, people harvest the grasses or cultivate annual crops in these tree-lined strips of public land.

AGROFORESTRY PRACTICES IN FORESTS, WOODLANDS AND WOODLOTS

Forest enrichment can make wooded areas more useful by protecting and improving soil and water resources, by increasing the production of tree products or by adding new productive plants and animals. This forest enrichment may involve only the selective cutting and protection of existing forest plants or it might also extend to the introduction of multipurpose trees, herbaceous crops and/or livestock. Where trees are planted to prevent or reverse erosion in forest clearings, they may be combined with soil and water conservation structures as well as herbaceous plants for ground cover.

The *taungya* agroforestry system combines the establishment of new forest plantations with food and cash crops. Farmers clear and prepare a site, plant their crops along with tree seedlings and maintain both trees and crops for a few years until the tree canopy begins to close. They then repeat the process in a new site. This approach can reduce the cost of reforestation, but in most cases farmers and forest dwellers work for low wages and eventually lose access to agricultural land and forest products. *Taungya* systems can, however, incorporate secure terms of use and access for rural communities to pursue farming as well as forest development over the long term. While this system has been used most widely in humid areas, it might also be applied in dryland Africa for the rehabilitation of grazing lands or the establishment of woodlots. It could be useful in any situation where deforestation and land degradation can be treated by a mixture of temporary cropping with tree establishment.

Woodlot enrichment is similar to forest enrichment, although it usually involves more intensive management of trees and other plants

in a smaller area. Permanent woodlots may be sited almost anywhere in the landscape, from cropland to pastures, but once an area becomes a woodlot it is a landscape feature in its own right with many qualities similar to forests. Existing woodlots may be enriched by introducing multipurpose trees, herbaceous crops or animals. New agroforestry woodlots may be designed to produce fuelwood and fodder, provide a more diverse mix of products and services and sustain the soil and water resources of the site.

Tree-crop plantations do not necessarily include agroforestry practices, but may do so. Herbaceous groundcover crops can be introduced into stands of commercial tree crops along with upperstorey trees used to produce shade, mulch or wood. For example, some farmers combine citrus and coconut groves with planted pastures subjected to controlled grazing. In other cases, coffee and tea plantations benefit from dispersed shade trees that improve soil fertility and provide fuelwood and timber.

1.2 The Setting: Dryland Africa

Subsaharan Africa is home to roughly 450 million people, living in a variety of physical, cultural and economic environments. The major ecological zones include lush tropical forests, cool grasslands on the high plateaus, wet montane forests and semi-arid savannah lands. These are depicted in three maps, showing climate, rainfall and vegetation zones for the continent, and one map, showing climatic zones in subsaharan West Africa.

The terminology given in the map of vegetation zones will be used throughout this book. This may be compared with two other widely used terminologies on the basis of average annual rainfall as follows:

Average Annual Rainfall (mm)	2500	2000	1500	1000	500	100	0
Vegetation Zones		Mosaic Woodland	Wooded Savannah	Shrub Savannah	Tree Steppe	Grass Steppe	
Francophone Terms (Aubreville)			Sudano-Guinean	Sudanian	South Sahel	North Sahel	Sahara
Anglophone Terms (Nigeria)			Derived Savannah	Guinea Savannah	Sudan Savannah	Sahel Savannah	

Within each of these settings, people's settlement patterns, styles of living and land-use systems may vary even more than the land itself. Most of the people throughout the continent are living through a period of rapid and dramatic changes in land-use patterns, economic conditions and the natural environment. The pace of change often exceeds the capacity of local organizations and national institutions to develop new land-use practices that support both the natural resources and the welfare of the people. This is especially true in the drier, more fragile zones where drought and famine have become increasingly common.

Savannahs and grasslands cover approximately two thirds of sub-saharan Africa and support a large part of the population. Crop and livestock production in this zone is increasingly limited by erratic and

Climatic zones in Africa.

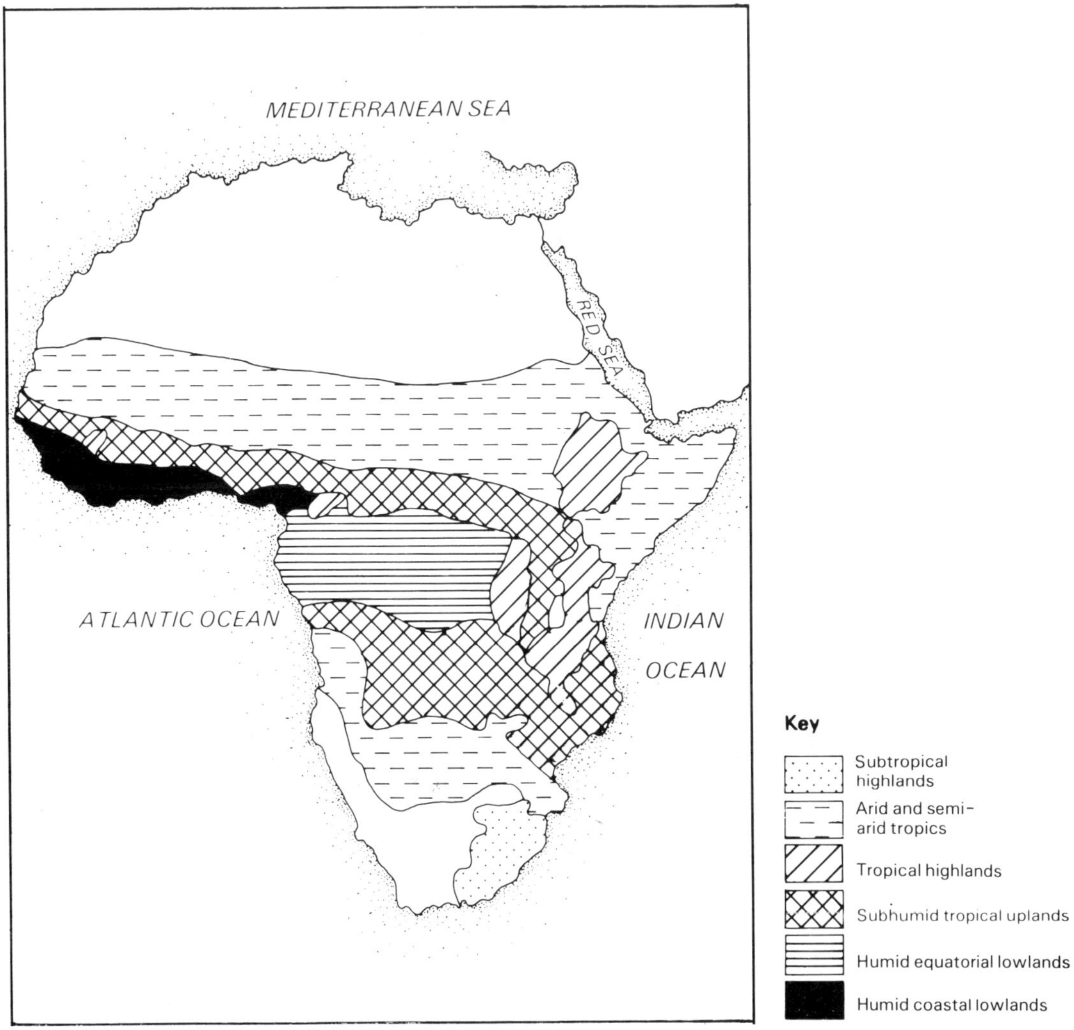

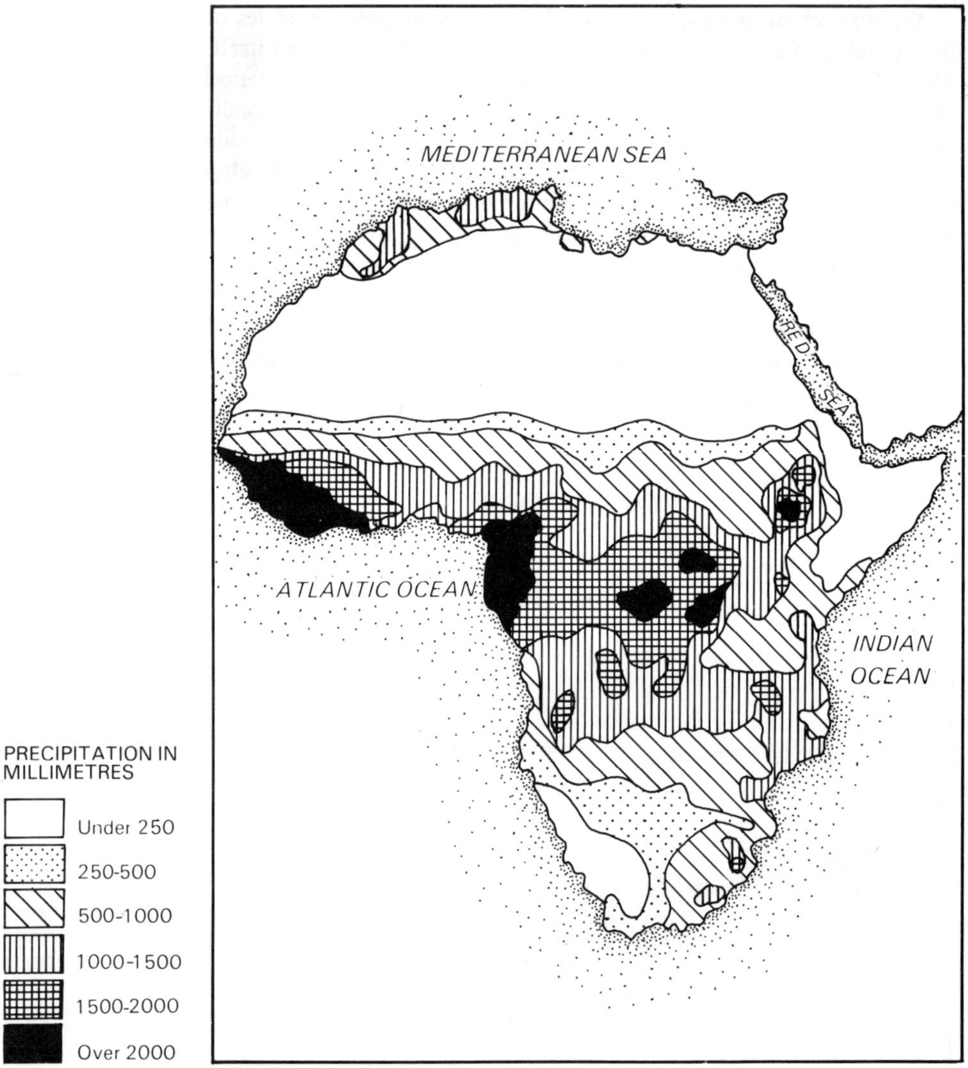

Rainfall zones in Africa.

insufficient rainfall and by the deterioration of soil, water and plant resources, a process often referred to as desertification. Overall, the trend is towards expansion of the desert and corresponding expansion of dry savannah areas at the expense of the moister savannahs and woodlands. The extent of this process is shown on the map of African ecological zones, with large areas already subject to some degree of desertification and others soon to be affected.

Desertification takes many forms and affects the lives of rural people in immediate, practical ways. The disappearance of vegetation, erosion of soils by wind and water and decreasing soil productivity—all these affect the livelihoods of herders, farmers, gatherers, artisans and traders. Under such circumstances, traditional land-management systems often give way to the pressure to produce enough to eat. People may reluctantly adopt farming, herding or tree-harvesting practices which will eventually destroy their natural resources or they may choose to move to another place.

People may also resign themselves to a lower standard of living and a more difficult existence than they have known in the past. For example, women who gather firewood may find that the best species of trees and shrubs are no longer reproducing themselves and are in ever shorter supply. This may mean less firewood for cooking and less frequent or less nutritious meals than they have had in the past. It may also lead to the use of animal dung for fuel, thus depriving the soil of an important source of nutrients in farming systems where dung would otherwise be applied to cropland. These same women and their children may also spend increasingly long hours searching for firewood.

Men, women and children who depend on livestock for their food and income may find that they must travel more often and over longer distances to find enough fodder for their animals. In some cases, the large trees which bear nutritious pods during the dry season are cut down by farmers or charcoal makers and herders have to buy dry-season fodder at the market or sell their animals at low prices or watch them die. As the trees disappear from the savannah, the soils beneath are exposed and in many cases the surface bakes to a hard crust, impenetrable to rainfall. The streams and wells, formerly fed by slow runoff or deep percolation of water, become less reliable in the dry season, forcing the herders to travel longer distances to water points.

Woman bent under a heavy load of fuelwood.

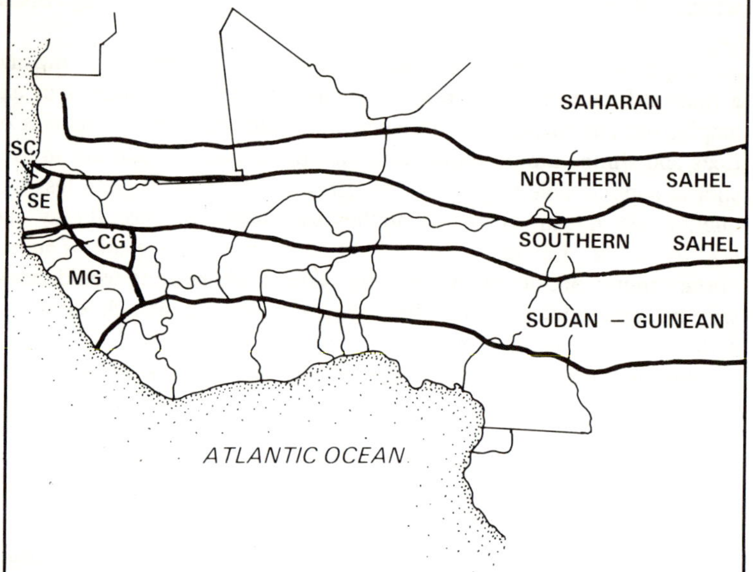

	Mean Annual Precipitation (mm)	Mean Annual Saturation Deficit (mm)
Saharan	Less than 200	20
Northern Sahel	200 – 400	15
Senegal Coastal Sahel (SC)	400 – 500	5.3 – 7
Senegal Sahel (SE)	500 – 900	9 – 12
Southern Sahel	400 – 1200	11.5 – 22
Sudan – Guinean	950 – 1750	7 – 12
Cosamance Guinean (CG)	1200 – 1750	6.5 – 7
Coastal Guinean (MG)	1950 – 4500	4.4 – 5.5

Climatic zones of subsaharan West Africa.

Whether they rely entirely on farming or keep livestock as well, people who farm small plots in the drylands are also severely affected by desertification. In areas affected by erosion, desiccating winds and dwindling water resources, entire crops may be lost. Even in better years, crop yields are reduced due to the low fertility and poor physical condition of eroded soils. As yields decrease, farmers often expand their cropland at the expense of grazing lands, woodlands or fallows. As this process continues, the shrinking sources of fuelwood, fodder, wild foods and building poles are threatened by overuse.

There is considerable controversy over the extent to which land-use practices are responsible for desertification or whether a large-scale climatic change is taking place. Even among those who believe that land use is the major cause of land degradation, there is wide disagreement about whether the problem stems primarily from land use by rural people or from national economic policy. In any case, it is clear that large numbers of people and vast areas of land are at risk. Land-use systems are needed that can adapt to the adverse effects of climatic change and prevent or reverse the avoidable damage caused by inappropriate use or overuse of resources. This implies an ambitious initiative to identify, revive, adapt or develop sustainable production systems for widespread use throughout the dry regions of Africa.

Any new land-use system must also be appropriate for the economic and social conditions of rural communities. It must provide for people's basic needs without requiring equipment or material that they cannot afford. In short, land-use systems are needed that provide useful products, that conserve and restore natural resources and that build self-reliance rather than dependence on expensive materials.

Women in a field ravaged by drought.

Any land-use system that meets these three criteria will be based largely on the traditions, knowledge, skills and ongoing experimentation of rural communities. In most of the dry areas of Africa, agroforestry practices have always been an important component of local land-use and resource-management systems. If the people living in these areas have been both victims and perpetrators of environmental degradation and desertification, they have also been wise users and healers of the land. Their experience and skills, varied as the environments and cultures of the continent, offer the practical beginnings for the development of ecologically sound land-use systems and hope for the future.

1.3 The Role of Agroforestry

Agroforestry is frequently invoked as a solution to problems of land and water degradation as well as an answer to shortages of food, fuelwood, cash income, animal fodder and building materials in subsaharan Africa. What can agroforestry actually contribute to the development of sustainable land-use systems in the region? In answering this question, it should be kept in mind that agroforestry is only one of several approaches for improving land use in any given situation. Yet the breadth and variey of agroforestry systems and practices imply that agroforestry offers at least partial solutions to many rural land-use and production problems. The promise of nitrogen-fixing trees for improving soil fertility in cropland and pastures has been widely discussed, as has the resistance of some trees to drought, the role of windbreaks in protecting croplands and pastures, the contribution of high-protein tree fodder to livestock production and the commercial potential of several kinds of tree crop. In addition, agroforestry practices are appropriate for a wide variety of places within the landscape, not just for cropland or pastures.

The droughts and famines of recent decades have alerted many to the need for rehabilitating degraded land and water resources in Africa's dry regions, as well as for developing appropriate, sustainable

Cattle on degraded pasture land.

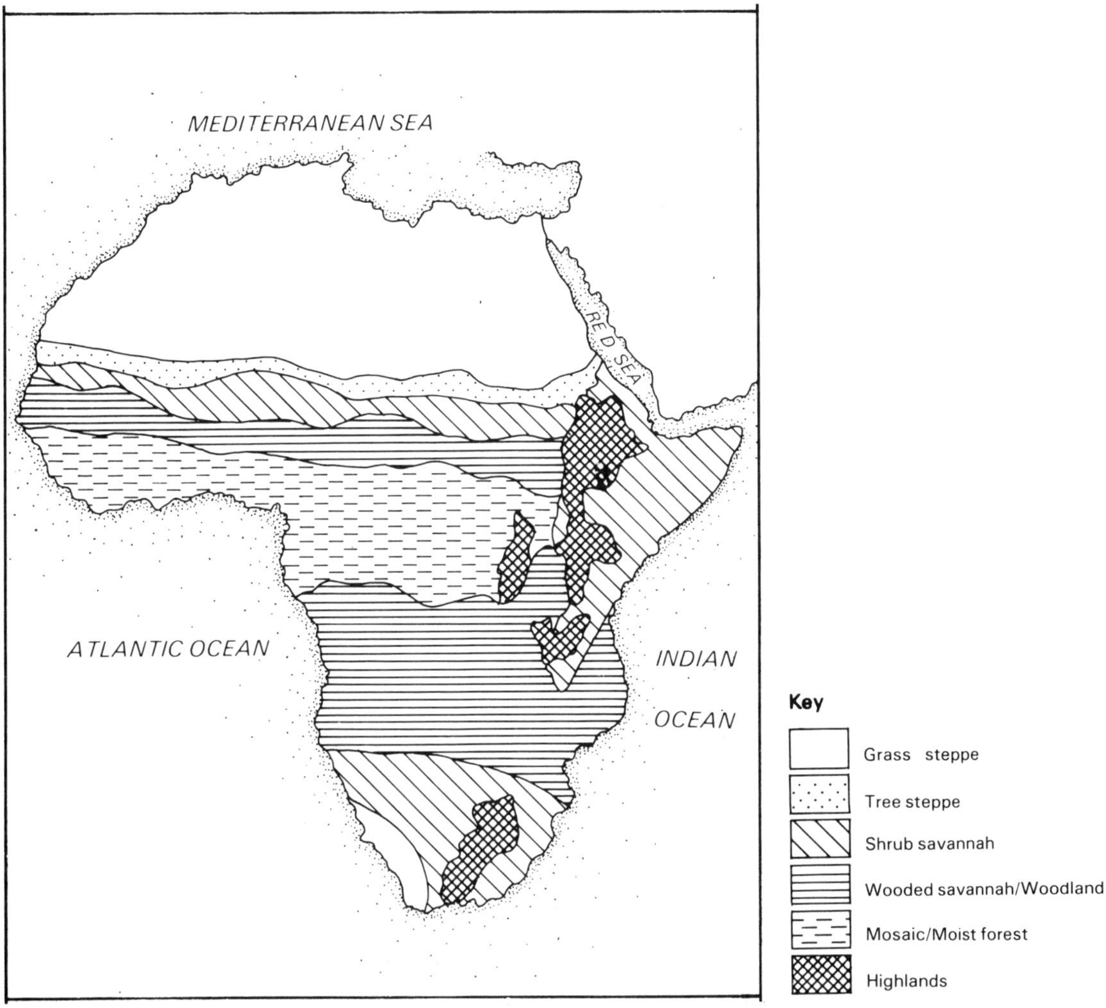

Vegetation zones in Africa.

land-use and production systems. Farmers and livestock owners in these areas need alternatives to the modern agricultural technologies and simplified cropping systems that have been promoted over the last few decades as part of development efforts. While the experts may disagree on whether famines, droughts and resource degradation are natural disasters or caused by misuse of the environment, there is a general consensus that future land-use systems and technologies must give people more flexibility to respond to rapid shifts in economic and ecological conditions. In addition, new production systems must maintain, or in many cases restore, the soil and water resources upon which rural life depends.

In this context, several traditional agroforestry systems have in fact sustained people for generations in a variety of African environments.

The intercropping of *Acacia albida* with millet and sorghum in the West African Sahel is one of the best known examples of a successful traditional agroforestry practice. Less well known but equally significant are the sylvopastoral systems developed by people who depend on managing livestock and their fodder sources in African savannah lands. At present, many of these successful traditional systems are literally losing ground in the face of disruptive changes and increasing pressures on the people and their natural resources. The challenge is to maintain those agroforestry systems which are now under threat and to improve and adapt long-standing practices which must now be effective under changing conditions. Where traditional agroforestry systems have not been used or can no longer be used because of changing conditions, new systems need to be developed.

Rural development, resource management and agricultural programmes have often provided separate, even conflicting, advice about how to use soil, water and plant resources. However, more recently a growing number of specialists in agriculture, livestock production, forestry, soil and water conservation and rural development have become aware of the interconnections among all these activities in the rural landscape. Most farmers and livestock owners have always been aware of these links because they must deal with many aspects of resource management in order to maintain their lands and livelihoods. Agroforestry offers a practical way to apply a variety of specialized knowledge and skills to the development of sustainable rural production systems. This is especially important in difficult environments, where people must manage hillslopes, dry farmlands and fragile rangelands to survive and earn their livelihood.

Fieldworker learning from local specialists.

1.4 A Land User's Perspective

Once the full range of problems and practices in a local area has been investigated, there will undoubtedly be several possible agroforestry alternatives to consider. However, there is a distinct lack of ready-to-use agroforestry information suitable for the diverse environments and circumstances of the people in dryland Africa. One response to this situation has been for researchers and development agencies to choose a few practices and a short list of species for testing under a variety of circumstances. Yet there is a shortage of time and resources for such a trial-and-error approach to agroforestry research.

Trees require a lot of space and a long time to grow, which limits the number of formal experiments and replications possible in a given time and place. This limitation implies a need for careful selection of the agroforestry species and practices to be tested by experimentation. In addition, it is unlikely that a programme of formal testing will result in land-use systems widely applicable to the numerous, distinct environments and cultures of dryland Africa. There is thus a clear need for a large number of less formal, community-based research and innovation efforts.

Agroforestry research and development workers in the field must carefully mix existing land-use practices with the science of designing and testing new practices. They must combine research, extension and evaluation in a way that is not often found in soil and water conservation, forestry, or agricultural development programmes.

Agroforestry is a practical, but complex science. It can be applied to single plots, to large tracts of land or to entire land-use systems. Agroforestry practices serve many purposes and supply many products to a wide variety of land users. In addition, agroforestry is usually introduced in situations where both production and natural-resource conservation are important. Agroforestry also often involves a broader range of activities than many agricultural or pastoral land-use systems. Approaches based on one product, to serve specific 'target' groups with transfer of 'technology packages' onto uniform blocks of land, are not likely to result in the spread of agroforestry practices useful to all members of a rural community.

To address the needs and concerns of the majority of land users, agroforestry research and development workers must deal with multiple uses, multiple users, the entire rural landscape in all its complexity and a variety of indigenous practices and technical knowledge. In all instances, field workers must also work closely with local land users as clients and co-workers in a spirit of service and cooperation. These are the elements of a community-based research and development approach to the design and improvement of agroforestry practices. While the concept may seem simple, each one of these elements entails a complex set of issues.

MULTIPLE USES

Agroforestry practices contribute a wide range of goods and services to the rural community. Trees may provide food, shelter, energy, medicine, cash income, raw materials for crafts, savings and investments and resources to meet social obligations. Trees used in agroforestry systems can also provide a variety of services, such as the improvement of soil fertility for crop production, the improvement of microclimate for crop growth and the control of crop pests. In addition, agroforestry practices are often designed to protect and improve the quality of natural resources—including soil, water, vegetation and wildlife—and to substitute for the destructive use of special environments, such as riverine forests, hillslopes and fragile rangelands.

To take full account of all these products and services requires a more complex and flexible approach than single-purpose techniques for the improvement of forestry, cropping systems or soil and water conservation. Agroforestry practices will also be carried out by a variety of land users who often have different priorities and use the same places and products in different ways. Some of the purposes and products associated with agroforestry are listed below:

CASH INCOME

- Employment (cash earnings)
- Sale of products (cash earnings)
- Substitution of own products for purchased items (less cash spent)
- Exchange of products for other goods (less cash spent)

FOOD SUPPLY

- Increased amount of food
- Year-round supply of food
- Better-quality food (nutrition, taste, easy preparation)

ENERGY SUPPLY

- Increased fuelwood supply
- Better-quality fuelwood
- Cheaper or more convenient fuelwood sources

SHELTER/STRUCTURES

- Building material
- Shade

- Protection from wind
- Protection from animals
- Definition of boundaries

SAVINGS AND INVESTMENT

- New forms of savings and investment (tree crops, orchards, tree products, agro-businesses, woodlots, improved pastures)
- Better profitability or security of existing savings and investment

MEDICINE

- Preventive (to maintain health)
- Curative (to treat diseases or injuries)
- Veterinary medicine

RAW MATERIALS FOR CRAFTS AND COTTAGE INDUSTRY

- Increased supply of materials
- Secure future supply of materials
- New types of material

RESOURCES TO MEET SOCIAL OBLIGATIONS

- New source of support for social obligations
- Improved existing sources of support

CONSERVATION OF SOIL, WATER AND PLANT RESOURCES

Water

- Increased amount of water for plant growth, domestic use and livestock
- Improved seasonal availability of water
- Improved amount, quality or timing of water delivery to dams and large-scale water works

Soil

- Protection from erosion, loss of nutrients
- Restoration of degraded soils
- Improvement of soil moisture and fertility

Vegetation

- Maintain or increase diversity of species and habitats
- Substitute cultivated tree products for overharvesting of fuel-wood, fodder and other items from unique or fragile woodlands
- Increase yield of useful products
- Improve conditions for natural regeneration of most desirable species.

MULTIPLE USERS

The identity of the group or groups that will be involved as participants, beneficiaries or victims of changes in land use and management is often overlooked. Most agricultural development projects have been directed at 'target groups' of farm owners and managers, while forestry projects tend to treat whole communities as a single participant. The reality of rural life is more complex than either of these approaches implies. Any agroforestry programme that is to serve the majority of the people must deal with a wide range of land users, many of whom are neither owners nor managers of farms.

The term 'land user' refers to every person who makes use of a particular place and its resources, including soil, water, vegetation and wildlife. In any location, most land users can be classified by the kind of activity they pursue, by their terms of access to land and resources and by the way they are grouped in respect to their use of the land. Agroforestry workers need to involve and serve several distinct groups as clients. Depending on the place and the land-use system, these groups might include paid farm workers, unpaid family workers, home managers, processors, traders and consumers, as well as 'farmers', heads of farm households and the landless. Herders and gatherers would also be clients, as would tenant farmers, labour-exchange groups or grazing associations using private, communal or government land. Individuals, household and family groups, self-help groups and cooperatives could all be treated as users, depending upon how people are grouped in carrying out agroforestry practices, as well as in enjoying the benefits.

Land users may be classified in different ways, as follows:

LAND USERS ACCORDING TO ACTIVITY

- Direct land users: gatherers, hunters, herders, farmers (largeholders, smallholders and gardeners), farmworkers (including unpaid family workers)
- Indirect land users: processors, market vendors, consumers

LAND USERS ACCORDING TO TERMS OF ACCESS AND OWNERSHIP

- Owners
- Tenants (paying fixed rent)
- Users by permission or exchange agreement (continuous, regular, occasional)
- Illegal occupants and users (squatters, 'poachers')

LAND USERS ACCORDING TO THE SIZE AND TYPE OF GROUP

- Individuals, differing by sex, age, marital status

- Households, differing by size, age, wealth, ethnic group, headed by man or woman
- Communities and community groups: villages, small settlements, extended families, clans, self-help groups, religious groups
- Companies, cooperatives or associations: large commercial enterprises, small businesses, producers' cooperatives, marketing cooperatives, farmers' associations, religious groups, ethnic groups

GOVERNMENT ADMINISTRATIVE UNITS

- National, provincial, district or local.

In any given place, land users and managers hold a variety of complex and ambiguous rights to land, trees and water. Rights of access to trees, in particular, have often been overlooked in land-tenure reform programmes, leaving conflicts to be resolved locally, sometimes through reference to customary law. As a result, the rights of individuals or communities to trees and tree products are often less clear than their rights to land. Rules of use and access must be understood and agreed upon as a basis for the development of agroforestry practices.

AGROFORESTRY AND THE RURAL LANDSCAPE

Many rural development and conservation efforts appear to be aimed exclusively at 'farms', cropland plots or blocks of forest or rangeland. By contrast, most people in rural areas depend on a complex pattern of land, water, vegetation, houses, roads, markets and public places. The arrangement of these features in the larger landscape reflects the history of the people and the land. For example, the place of trees in the existing landscape indicates a great deal about past tree planting, management and use. The patterns of land use and settlement can also indicate how land, water and trees are owned or controlled and whether they are managed by large groups of people, small groups or individuals.

People's ideas about the form and shape of their surroundings and the functions of specific places may help to determine which land-use practices are appropriate for particular sites. Agroforestry is based on the careful placement of trees in spaces shared with crops or livestock: many rural people are also concerned about the fit of an agroforestry practice into the larger landscape, judging the result in part according to their image of what a particular space is for and how it should look. By learning more about why land, water, plants, roads and settlements are arranged in a particular way, agroforestry workers may be better able to help choose the right trees for the right places.

There is another practical reason why the visible landscape should serve as the point of departure for any development of a new or improved agroforestry practice. To plan agroforestry practices exclusively for blocks of cropland, rangeland, woodland or woodlots would mean missing the rich opportunities presented by boundary lines, roadsides, riversides, home compounds, public meeting places and all the in-between places where trees can fit into spaces not already allocated to crops, pasture or other uses. Such spaces can often accommodate trees planted individually, in lines, in clumps or irregularly dispersed according to patterns of soil, drainage, settlement, ownership or other factors of local importance.

The many facets of the rural landscape are not just an opportunity: they are also the visible result of a process of community development and land-use change. Most of the marginal lands in dryland Africa are experiencing rapid, sweeping changes. Large tracts of woodland and savannah are being converted to cropland and any landscape—whether forested, treeless or interspersed with trees—may change dramatically over one generation. Under such conditions, any agroforestry practice introduced and adopted on a large scale is bound to affect and to be affected by changes in land use and the evolution of the local landscape.

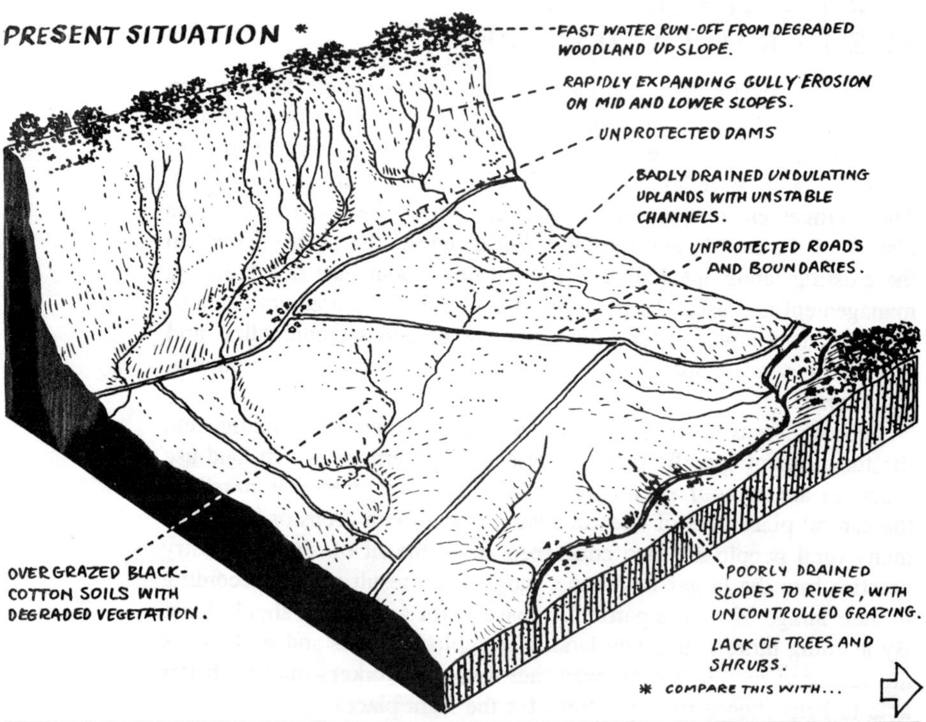

Since agroforestry practices almost always involve long-lasting, visible changes in the landscape, they must fit into future patterns of land use as well as into the present. Agroforestry can help to transform a treeless, barren landscape into one with trees closely knit into croplands, boundary areas, home compounds and grazing lands. Agroforestry may also be used to change a forested area into a planned settlement, with a mixture of forest, trees dispersed in pastures and multistorey cropping systems. Trees may be planted in blocks alternating with blocks of cropland or closely intercropped or in clumps or lines to fit between croplands, along roads, in home compounds or in public places. The choice must rest with the rural communities whose futures are affected.

People are rarely indifferent about the future of their surroundings. They usually have an idea about how they would like it to be and what it is likely to become. Their plans, expectations and hopes for the future can provide a map to guide the development and placement of agroforestry practices.

Where people already have a strong sense of how the landscape is developing and where they are happy with current trends, then agroforestry practices must be designed to be compatible with other changes in the landscape. For example, if people expect all grazing

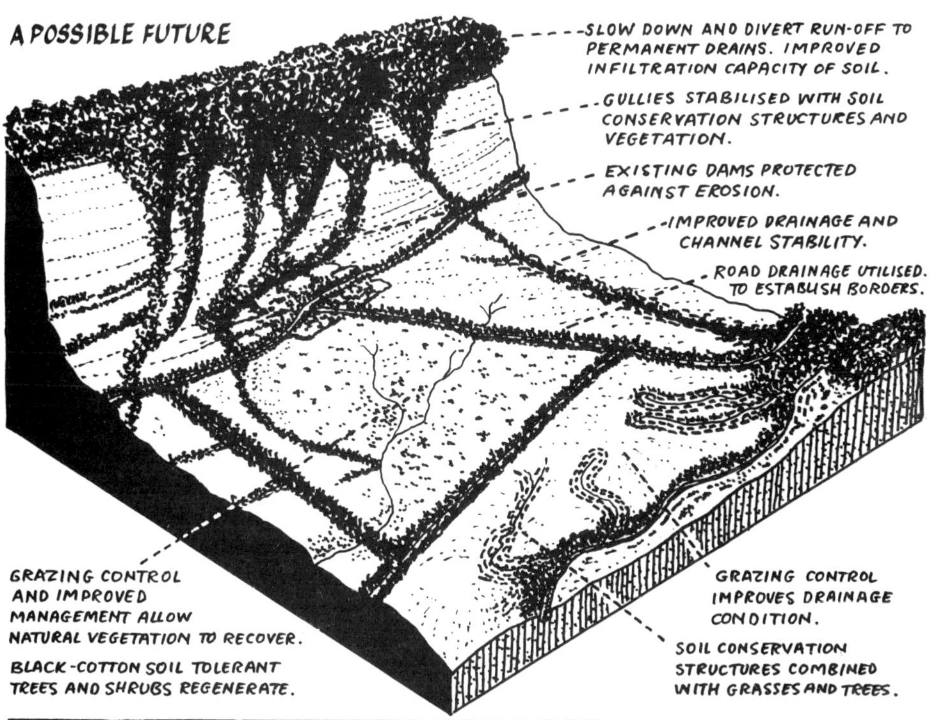

lands to be converted to cropland within the next 20 years, then the design of any sylvopastoral system should include only tree species and spacings that could eventually be compatible with cropping systems or could be harvested before the site is converted to cropland.

Agroforestry workers can also influence the shape of the future landscape by presenting new alternatives that may broaden the range of possibilities for communities to consider. With new alternatives, it may be possible to avoid problems that seemed inevitable or to achieve goals that seemed out of reach.

In conclusion, agroforestry practices are always an integral part of the larger landscape. Field workers must design practices that are compatible with current trends or that will help bring about alternatives chosen by the local community. To ignore this is to impose unknown changes on the future landscape without being held accountable to the people whose lives will be affected.

INDIGENOUS TECHNICAL KNOWLEDGE

The people who live in an area and use its resources possess valuable knowledge about the land and its uses. Field workers can develop successful agroforestry systems if they are able to learn from and improve upon indigenous knowledge and practices, combining their own efforts with those of local experimenters.

The scientific community knows little about the trees, shrubs and wild herbs that people use in dryland areas of Africa. Researchers need to work closely with land users to identify promising species for agroforestry systems and to understand what the local people already know about the environment and the local economy. Such information includes the interactions of trees with soils, crops, pests and diseases, as well as their uses, management and ownership.

Where an agroforestry system is already well established, some people will know how to select, prepare and manage a site and how to select, breed and cultivate the trees and other plants used. This knowledge may be based on observations of various plant combinations and spacing arrangements or on active experimentation. People who have long experience with indigenous agroforestry systems have a wealth of knowledge about plant, soil and water interactions, as well as pest management, processing and use of products and often marketing. Sometimes such knowledge is formally recorded and passed on. In other cases, the knowledge of the system and how it works is bound up in the practice itself. Such knowledge and skill may be widespread throughout the community or concentrated among local specialists.

Researchers and development workers can learn from the local community by studying their formalized knowledge about agroforestry systems or by observing and recording actual practices. Outsiders can also ask specific questions and measure selected aspects of agroforestry systems and their performance.

In places where agroforestry systems do not exist or are not well developed, people still have a detailed knowledge of their environment, including the most useful plant species found in forests, rangelands or farmlands. This detailed local knowledge of cultivated, wild and partially domesticated plants may be the key to selecting woody and non-woody species for new agroforestry systems. Often local people can identify useful species and source areas for high-quality plants and seeds. For any given species, they may know the plant habitat, growth rate, method of regeneration, compatibility with other plants and interaction with animals and insects. Often people have also observed the response of a species to changes in site conditions, such as fire, drought or floods, and responses to different management practices, such as lopping, pruning, reduction of shade, soil tillage or intercropping. Even where local plant species are not used directly in new agroforestry systems, local knowledge of the land and the natural

Women explaining the qualities of fruits.

vegetation can help in choosing new species and combinations which will be compatible with the site.

Rural people also have a great deal of useful information to contribute in their role as consumers. Even if they are not familiar with the production or ecology of a particular plant or group of plants, information about local preferences and uses of plant products can help agroforestry workers to identify the most appropriate species for introduction into an area or the most important qualities to seek through plant breeding.

Not only can people's past experience and traditional knowledge help to guide the development of new agroforestry systems, but their judgement and skill as experimenters in their own right can make a significant contribution. Local people can use their experience in innovation and evaluation to identify and develop useful new practices, and their full involvement can also help insure that new practices are widely adopted. Active local participation is important for improving and adapting traditional agroforestry systems and also for fitting trees into new situations, such as settlements, cropping systems or rangelands.

Local knowledge and experience also extends to systems of allocation, ownership and management of resources. In every community, there are mechanisms for formulating and enforcing rules and resolving conflicts concerning the use of land, water and trees. These mechanisms are a valuable component of indigenous knowledge, particularly relevant to the development and improvement of agroforestry practices.

LAND USERS AS CO-WORKERS AND CLIENTS

The active involvement of the local community in the development and testing of new practices can make the critical difference between the success or failure of an agroforestry development project. For a start, considering local people as clients means integrating their needs and concerns into every step of project planning and implementation. Yet clients may be passive recipients of services, rather than active participants.

Experience throughout Africa suggests that new land-use practices are more likely to succeed over the long term if land users participate actively in the entire process of problem definition, design and testing of solutions and extension of result. In this case, participation means

The participatory process in action.

far more than the provision of labour. A fully participatory approach incorporates the skills and judgement of local people acting as partners throughout the research and development process.

Community-based participation also requires that all categories of land users be treated as clients. This will usually mean dealing with two or more distinct client groups and may involve conflicts of interest between them. If land users are actively participating in the planning and implementation of agroforestry projects, then they may also take the lead in resolving conflicts of interest among themselves, with outsiders providing information and ideas as needed. In some cases, disadvantaged groups may ask for direct assistance with mediation in order to assure that such conflicts are resolved fairly.

Community-based research and development workers who follow the approach outlined here can contribute more to a local community than a few new agroforestry practices. They can help to build local skills for continuing analysis, design and management of agroforestry and related land-use systems to ensure sustainable productivity for future generations.

SELECTED REFERENCES

Food and Agriculture Organization of the United Nations (FAO) (1986). *Tree growing by rural people.* Rome: FAO, 130pp.

Fortmann, L. and Bruce, J., eds. (1988). *Whose trees?* Boulder, Colorado: Westview Press, 341pp.

Nair, P.K.R. (1987). *Agroforestry systems in major ecological zones of the tropics and subtropics.* Working Paper No. 47. Nairobi, ICRAF, 15pp.

Raintree, J., ed. (1987b). *Land, trees and tenure: proceedings of an international workshop on tenure issues in agroforestry.* Nairobi: ICRAF and Madison, Wisconsin: Land Tenure Center, 412pp.

Richards, P. (1985). *Indigenous agricultural revolution: ecology and food production in West Africa.* London: Hutchinson, 192pp.

Rocheleau, D.E. (1987). The user perspective and the agroforestry research and action agenda. In H.L. Gholz, ed. *Agroforestry:realities, possibilities and potentials.* Dordrecht: Martinus Nijhoff and ICRAF, pp. 59-87.

Rocheleau, D.E. and van den Hoek, A. (1984). *The application of ecosystems and landscape analysis in agroforestry diagnosis and* *design: a case study from Kathama Sublocation, Machakos District, Kenya*. Working Paper No. 11. Nairobi: ICRAF, 92pp.

Weber, F. and Hoskins, M. (1983a). *Agroforestry in the Sahel*. Blacks-burg, Virginia, USA: Virginia Polytechnic Institute, 102pp.

Weber, F. and Hoskins, M. (1983b). *Soil conservation technical sheets*. Moscow, Idaho: University of Idaho, 124pp.

Weber, F. with Storey, C. (1986). *Reforestation in arid lands*. Arlington, Virginia, USA: Volunteers for International Technical Assistance (VITA), 335pp.

CHAPTER TWO

PARTICIPATORY PLANNING: PROCESS AND METHODS

2.1 Introduction

The approach taken by this manual relies on learning from and working with rural people to develop land-use systems suited to local needs. It suggests a general participatory method for the selection of agroforestry practices, adapted to specific needs and conditions. In some cases, this process may help to focus formal research activities more clearly. In most cases, it will lead to innovations and testing by farmers, development workers and community-based researchers.

In order to develop appropriate agroforestry systems, it is important that researchers and development workers have a clear sense of the local community, the physical environment, what skills and knowledge are available and how the natural-resource and production systems work. They should also know what needs are greatest in the community, what objectives and priorities people have for their land-use systems and what resources they have for reaching their goals.

This process usually begins with a review of the field workers' own goals and resources, the regional context of the project and the information already available about the site. Depending on their primary objectives, researchers, development workers or extension agents might use this book in very different ways. A project focusing on soil and water conservation, rural development, agriculture or forestry might also require that some kinds of questions be emphasized and others be omitted.

The maps of African climate, vegetation and rainfall zones in Chapter 1 may be used to put an agroforestry project into a regional context. These maps are related to the information on species and practices presented in Part II and Appendix I. The location of a project within a given ecological zone will also help to identify the most relevant sources of information among the projects and contacts listed in Appendix VI. These contacts are provided in order to encourage visits and the exchange of information, experience and seed among groups working in similar environmental or socioeconomic settings.

The wide range of information required could imply several years of formal surveys and studies. What is suggested instead is to hold a series of broad-ranging, informal discussions with members of the community, both individually and in groups. These discussions may involve different kinds of people in a variety of situations. For example, a field worker might attend a meeting of a farmers' association, convene a meeting of the community, meet with women at water points or with women's labour-exchange groups as they work, accompany people on short gathering or herding trips, talk with people in their homes or fields or interview different individuals from the same household.

Partnership implies a two-way flow of questions and information, with participating community members also learning from the field workers. Everyone involved should gain a shared understanding of the needs and interests of the different land-user groups within the community and an accurate picture of the existing and potential uses of land, water, plants and their products.

Records of these discussions may have many other uses beyond agroforestry planning. They may serve as background references for local residents and future community workers and may be used to design learning materials for schools or agroforestry training programmes. These surveys are also likely to stimulate discussion on land-use activities other than agroforestry.

This chapter presents several methods for exploring land-use issues with rural communities. These include a variety of informal 'rapid-appraisal' surveys based on direct observation, discussion and participation in the field. The goal is to describe present land-use practices and systems and to formulate practical improvements. The interpretation and application of survey results depend on the experience and judgement of field workers and community members, rather than on formal statistical analysis.

This first stage can be very informative, even if field workers are already familiar with the local situation. By showing an interest in new topics and asking for peoples' opinions and suggestions, field workers may uncover a whole new range of knowledge not available to them before. The surveys also provide an opportunity to meet groups that may not have participated in agroforestry activities before, such as women's groups, farmworkers, herders, herbalists or charcoal makers.

There are several ways to learn about traditional knowledge, current practice, trends in land use and management and local experimental initiatives. The learning surveys described here include:

1. A review and summary of what is already known by participating research and development workers
2. General field visits, or rapid surveys, to describe the landscape, land use and land users
3. General and specialized group interviews to identify land-user groups, land-use practices and systems, local knowledge and land-use changes, problems and potential solutions
4. Household interviews to discuss the same topics
5. Walking interviews with individuals and small groups to discuss the landscape, land uses and activities

6. Individual interviews focussing on specialized knowledge and skills or discussing topics which are less suitable for group discussion
7. Action interviews (participant observation), which involve working with individuals or small groups to learn from direct experience about their land-use systems and activities.

It is usually best to combine several approaches to get a reasonable idea of what is common knowledge and what information can be contributed by specialists. Several types of specialized expertise may be available in the community—from herbal medicine to fruit-tree horticulture—which will help field workers to understand the environment as a whole and to identify many useful plants and practices. It is also important to understand what are common knowledge and practices and to what extent community members have developed and experimented with new ideas. This information can give field workers and community members an idea of how to present new information, how quickly to introduce new agroforestry practices, which kinds of practices to try first and who would be most interested in trying them.

Usually a few individual interviews and group discussions will provide a wealth of information and ideas. Group and household discussions may involve several visits to different households and groups. Specialists are often interviewed in detail on an individual basis and members of special groups, such as women, landless people, farm workers or minority groups, may also be interviewed individually to discuss their views and interests in a private, informal atmosphere.

Several types of information may be obtained from these interviews and discussions. These may be summarized as follows: (1) preliminary information on the people, their environment, how they see it and how they use it, (2) current practices in the management of land, water, crops, livestock and wild plants and animals, (3) needs and future plans, (4) land-use problems and potential solutions, (5) changes in land use, (6) specialized knowledge concerning the management and use of land, water, plants and animals and (7) experience and future plans related to agroforestry and other sustainable production systems.

Examples of information-gathering techniques and topics are given in this section to serve as general guidelines, not as fixed instructions. Field workers should make use of several approaches to learn about local community members and their use of land in their own terms. Some topics are best handled in individual interviews, others on walks through the landscape and others in work sessions with field workers joining community members in their tasks. Here, each survey technique will be described in relation to a particular set of information in order to present the major techniques and some of the most important topics in a single brief chapter.

Each of the methods and each of the special topics treated here could be recombined to suit a specific situation. For example, while it is true that group interviews are especially useful for obtaining general descriptions of land-use systems and problems, household interviews can also provide much of the same information, depending on the time available, the skills and preferences of the interviewers and the experiences and preferences of community members. As another example, in some places women speak out most freely in women's groups, in other places as individuals and in yet other cases they speak most openly within their families or households.

This manual suggests a general sequence of learning and discussion activities and describes several types of informal surveys and interviews, with a short list of questions for each. More detailed lists of questions and sample survey sheets are provided in Appendices III and IV. If this approach is successful, the collection and summary of information will continue and written records will expand for continuing reference by the community. Local information obtained from these surveys, backed up with the more general, regional information presented in Appendix I on agroforestry species and their uses, can help to decide which agroforestry practices and species to consider for introduction. Part II of this manual then provides guidelines on how to design the specific agroforestry practices chosen through this mutual learning and discussion process.

2.2 Initial Mapping and Information Summaries

Beyond a very general understanding of climate and vegetation regions, community-based agroforestry workers must have a practical knowledge of local physical and social conditions. Some critical questions can be answered by direct observation, while others can only be answered by the people living on or using the site (see Appendix III).

First of all, field workers will need to define the site. Is it a community, an administrative unit such as a village or division, a watershed, a group land grant, or a planned settlement? Within the boundaries of the site, they should know something of the physical surroundings, including the variations in slope, soil, drainage, natural vegetation and climate. The physical conditions for plant growth may differ dramatically within the same agroecological zone and even within the same site.

Woven into the physical environment are the land-use and settlement patterns that can transform similar physical surroundings into completely distinct landscapes. People and their dwellings may be concentrated in one place, in clusters or dispersed. Settlements may be associated with specific physical features such as hills or rivers, or with a particular type of land use. These patterns may have as much—or more—influence on the choice of agroforestry species and practices as climate or soil.

A realistic picture may be assembled by referring to previous knowledge or written descriptions of the site, to direct observations or to discussions with the people who live and work there. At this stage, if field workers know the place well or have good information from outside sources, such as maps and aerial photographs, it is useful to make a few rough sketches of the site. One sketch could include site boundaries, water points, streams, slope and type of terrain. A second sketch might show the distribution of major soil types. A third might be a rough map of the major land-cover types, such as forest, woodland, savannah, open grassland, annual crops, perennial crops and bare soil, and land uses, such as conservation reserves, gathering grounds*, areas used for grazing and browsing, croplands, public market and meeting places, homesteads and gardens.

It may also be useful to prepare a rough sketch of how an individual homestead, camp or cluster of homes might look, as well as a whole farm or a cluster of croplands. The place of pastures and woodlands in

*These are shared collecting areas that are sources of such products as wild foods, herbs, cut-and-carry fodder, fuelwood, timber, fibres, dyes and carving wood.

the larger landscape is especially important. Aside from the woodlands, where do trees and shrubs appear and how are they grouped?

One good reason for completing this exercise, even if the field workers are already familiar with the site, is to obtain a set of rough maps of the area and its resources that can later be compared with the maps and descriptions provided by the land users themselves. The maps and notes prepared at the outset will also provide a practical focus for future discussions with the community and will help field workers understand how community members see their own resources and how their perceptions differ from those of the field workers or the results of formal surveys. This comparison should help the field workers discuss local knowledge and practices and the choice of species, places and combinations of plants for new agroforestry systems. It may also be useful for developing mutual understanding in future discussions with members of the community.

These maps and summaries and the process of preparing them can give field workers a sense of what they know, their differences in approach and what they need to find out and agree on before they begin working more directly with the local community. This experience can also sharpen the observations made during the first field surveys.

2.3 Rapid Survey of the Landscape and Local Community

Field visits may confirm or change prior ideas and images of land use in an area. They will almost always add some new information, and field workers can note the differences between their earlier sketched impressions and what they actually observe (see Appendix III). One of the most important results is a better understanding of the field workers' biases and incorrect impressions. This knowledge can help them learn to observe actual field conditions more accurately.

It is usually best for outsiders to explore a project area in the company of a local community member. In addition to serving as a guide, a community member can help field workers avoid trespassing, invading peoples' privacy, or simply making people uncomfortable. Important contacts are made and lasting impressions are formed during this period of introduction (or reintroduction) to the people and the land.

Depending upon the terrain and the transport and time available, field workers may choose to spend more time walking, sketching from a high point in the landscape or driving through large tracts of similar terrain with occasional stops for closer observation of a particular farm, field or type of soil or plant. This can be a stimulating and informative activity for all involved if done in a small group that includes people from the area familiar with the local variability of soils, cropping systems, settlement patterns and social groups.

In order to understand the surroundings from different perspectives, it is advisable to repeat this activity with people of different ages, families, social groups and occupations. The goal is to understand the full range of conditions that exist in the area. It is especially important to be sure that the roads, paths and observation points covered provide a realistic and complete picture of the landscape.

It is also important to learn about local structures of decision making and authority in order to decide how best to introduce and organize agroforestry activities. Appropriate levels and modes of cooperation with local authorities and community groups should be explored carefully to avoid becoming too closely associated with any one group's special interests. It is important to work with a wide range of people in the community, not only the wealthy and powerful or the most vocal.

It is usually best to discuss survey plans with the appropriate community leaders before beginning the survey process, though it may or may not be appropriate to involve them directly in the first field visits. If a formal presentation to the community is customary, then such an occasion can provide an opportunity to introduce the participating field workers and the purpose of the surveys. Such events also allow field workers to observe the role of local leaders within the community: the patterns of authority and influence will become clearer later on through informal talks with a variety of people from different groups.

After a series of introductory visits, agroforestry research and development workers should be in a better position to plan more detailed discussions with individuals and groups. Following a thorough review of their information summaries and introductory visits, field workers should be able to adjust the style and content of interviews and meetings to suit the local situation. The more intensive period of survey activities usually consists of a combination of group, household and individual interviews, plus 'walking discussions' and direct participation in work sessions.

2.4 Group Interviews

Group interviews are one of the easiest ways to begin working with a local community on an agroforestry project. In a series of group interviews and discussions, field workers can explore the variety of local crop and livestock production systems and sources of livelihood, the community's perception of the environment, including land, water, plant and animal resources, past experience and present expectations concerning land-use and environmental changes, local uses of trees and wild plants and any land-use problems encountered. To cover all of these topics usually requires two or three visits with each group. In most cases, field workers should also follow up with individual interviews on specific topics. The results of these discussions can be summarized using the forms in Appendix III or any format convenient for local use. At the very least, these discussions should provide a basis for the choice of species and design of agroforestry practices.

It is important that all groups of land users are included in the discussions, not just owners annd managers. Often field workers will learn about new groups as they conduct household and individual interviews, and this will lead to subsequent rounds of group discussions.

The best way to approach a group for the first time depends on what the leaders and community expect in terms of formal introductions and whether the field workers are already known in the area. Where advance notice and formal introductions are required, local leaders may dominate initial discussions, in which case it is important to gather more information in less formal follow-up visits.

If it is possible, one of the simplest and most effective approaches is to meet with a group that is already together in a place of work or a public place. People may be willing to talk in this setting about local environmental and land-use systems or they may arrange a meeting at a more convenient time. Meetings may also be arranged by community members familiar with both the local groups and the field workers.

When they first meet a group, field workers should introduce themselves and their reasons for seeking local knowledge and opinions about land-use systems. It is also useful to ask groups to introduce themselves, including their purpose and activities.

Since agroforestry practices combine trees with animal and crop production and can fit almost anywhere in the landscape, it is important to discuss how the whole landscape and land-use system works, including the views, goals and expectations of various groups of people. If field workers begin by talking about trees, many people will assume that they are only interested in forestry. Participants may then fail to discuss the use and management of other land, water and plant resources that could be important for the design of future agroforestry systems. For example, they might neglect to mention wild fodder and food plants found in woodlands, pastures or fallow land which could be integrated into agroforestry systems. Also, most community members have encountered outside workers who are interested in trees only in terms of cash crops or forest conservation reserves, or who are only interested in crop agriculture or soil conservation, so it is important not to place too great an emphasis on any one of these topics.

One way to begin is to ask how long the community has lived in the area and whether the people present are from the area originally. If not, they can explain where they came from and when and why they moved (see Appendix III). In some cases, this will lead directly into discussions of land use, production systems and livelihoods.

OCCUPATIONS AND LAND-USE SYSTEMS

People usually respond readily to questions about local farming and land-use systems. What do people do to make a living? Some occupations may be widespread while others are limited to a few specialists. If relevant specialists are present, they can be identified during group discussions for possible follow-up interviews. Some people may be employed for wages, either outside the area or locally. Information on wage employment can help to clarify the division between employers and employees and give an indication of the availability of labour for new practices.

Those present at the discussion may also be involved in different kinds of land-use and production systems (see Appendix III). For example, there may be herders and farmers, further divided into goat herders and cattle herders or cash-crop farmers and subsistence farmers, each with different interests. Charcoal makers and timber harvesters, or small- and large-scale farmers might also constitute distinct groups with respect to agroforestry. As people discuss present livelihoods and forms of land use, they may progress into a discussion of land types, land-use history and the changing condition of natural resources.

LAND CLASSIFICATION

Whether they are recently settled or have a long history in an area, people have some practical way of classifying their surroundings into different categories. The system of classifying land uses and the natural environment may be formal or it may simply reflect widely held views about the production potential or management of specific elements of the landscape. If people from different groups have recently settled in the area, they may have different views. However, all residents will usually share some basic perceptions. If more than one group uses the site, it may be useful to meet with each group separately to learn how they view their surroundings.

Perhaps the best way to learn about systems of land classification is to ask a group how many different kinds of land there are in the area and what they are called. The word 'land' should be translated carefully so as not to mean soil or landform. Only very general terms should be used in order to learn what natural features are most important to the group. They may define land types by vegetation, soils, landforms, land use or a combination of these factors. Group members may be able to sketch or point out different land types and land uses on-the-spot. At this point, field workers may also identify individuals who are particularly knowledgeable about land types, vegetation and land use for individual interviews and mapping exercises.

The classification of land types and resources is important for any discussions of land-use problems and possible changes. While it will rarely be possible or practical to document traditional land-type or land-use classification systems precisely, it is essential to be able to discuss resource management, production problems and agroforestry practices in terms of specific classes of land and land use which are meaningful to the local community. For instance, a particular species or cultivation practice may be suitable for steep slopes but not for valley bottoms, while trees that do well on dry, sandy soils may not survive in nearby pockets of clay soil or along river banks. It is important to understand local land and land-use classification sufficiently to identify appropriate sites, species and management strategies for the introduction of agroforestry systems (see additional questions in Appendix III.)

LAND-USE HISTORY AND CHANGE

Land-use systems in most of Africa are changing rapidly and have been changing for some time. Local people can give outsiders a sense of what the area used to look like, how rapidly and in what ways it has changed and what their concerns and hopes are for the future development—or rehabilitation—of their surroundings.

In initial group discussions, people could be asked to describe in general terms what the landscape was like in the past, including the condition of a few key features such as forests, grasslands, croplands, water sources and settlements. Details can be filled in later through interviews with knowledgeable individuals (see Appendix III).

Usually people can indicate places where forests have been cleared, grasslands converted to croplands or croplands degraded, abandoned or converted to grazing lands. Most people also notice whether erosion is becoming more widespread or severe and they can often identify the link between specific erosion features and particular practices or land-use changes. This information will help determine whether which land and water degradation are due to poor land-use practices and can be halted or reversed by improvements in land use, including agroforestry.

Discussions with the local community about the history of the environment and its current condition can also indicate to what extent people are aware and concerned about land and water degradation and how they respond to these problems. They may be willing to change land-use systems and management practices to protect or restore soil and water resources or may already have done so. The prior experience of the local community with conservation practices, both voluntary and imposed, is also an important factor in deciding how to proceed with the development of agroforestry and related practices.

CROP AND LIVESTOCK PRODUCTION

A group discussion usually provides good information on which crops and domestic animals are produced in the area, in what combinations, in what kinds of fields or gardens and which are for sale or for home consumption. Often groups can provide enough information about the work required for crop and livestock production, who does what jobs and when for field workers to draw a rough activities calendar (see ex-

amples in Appendixes III and IV). Other important topics for discussion include the size and quality of land holdings and the nature of land rights, as well as crop rotation, herding, land clearing and soil- and water-management practices.

People may provide a great deal of information about crop and livestock management without needing to be prompted by direct questions. In the process of listing crop and livestock management activities and sketching out the activities calendar, the group may volunteer all the necessary details. However, as some people describe management systems, it is useful to check occasionally whether the whole group agrees, either by a show of hands, voices or nodding. If one person or a few people dominate the discussion, field workers may need occasionally to pose direct questions to others in order to ensure a balanced view (see Appendix III).

TREES AND OTHER PLANTS

It is important to find out how people and their livestock use trees and shrubs. Agroforestry programmes in the field may involve management and protection of existing trees, domestication of local species on farms or grazing lands, introduction of local species into new plant combinations, sites or spatial arrangements or introduction and testing of 'exotic' woody species from outside the area.

In any case, it is important to know what trees and shrubs are already used, how they are used, who uses them, who controls access to them, where they are located and how they are managed. Information about changes in any of these areas may indicate if favourite plants are disappearing, if access is becoming difficult, if old skills or sharing arrangements are being eroded or if preferences or markets are changing. Critical areas of concern may be identified where there is a widening gap between what woody plants people have and what they need or would like to have. Where such gaps are growing, people may be motivated to recover past conditions or to develop new agroforestry systems that provide the same products and services. In either case, it is essential to pay careful attention to past practice, the present state of resources and current uses and preferences with respect to particular plant species.

In a group discussion, it usually makes sense to ask how trees and other plants are used and then to ask which are the most important species for each use and why. To learn and record the local names of common plants, field workers should keep a list in a small notebook with spaces for local and Latin names that can be completed in the course of interviews and field work.

If too many species are mentioned, the field workers should ask people to rank the best and most important species for each use. The objective is to get an idea of the most popular and commonly used species and people's reasons for choosing them. People may also be able to identify favourite species that are in short supply. A more complete list, along with plant samples, should be obtained in specialized follow-up discussions with individuals or groups (see Appendix 3).

To apply this information to agroforestry and related practices, it is also essential to know where people are willing to plant or manage trees. One indicator is a list of the places where they have maintained or planted trees in the past. It is also helpful to find out whether there have been any changes in the types of spaces that trees occupy in the landscape. Sometimes people also mention the spaces where trees or shrubs might be increased or introduced where they have not been kept before (see Appendix III).

People can usually identify the best places for introducing or maintaining trees, but field workers should keep in mind which places in the landscape have not been mentioned. When the group has finished naming the places they consider appropriate for trees, field workers can ask about some of the other places and the reasons for not mentioning them. This can help distinguish the places where trees and

shrubs might be introduced with appropriate management from the places where they could not be grown under any circumstances.

Field workers also need to find out what local resources are available for raising, managing and planting trees, shrubs and grasses in agroforestry systems. They need to ask what people know about growing and managing trees and other plants, how much experience they have had with different species and in which environments.

The skills and physical resources necessary to produce tree seedlings may already be available in the community or district, either through private, government or group nurseries, seed suppliers or information centres. However, in most situations the community will have to pool their own information and skills, collect some of their own seeds or at least grow some of their own seedlings.

In addition to general group discussions, field workers may wish to conduct specialized group interviews. These might concentrate on trees and wild plants—their uses, ecology, site requirements, management, potential for domestication, planting niches and resources for propagation and management. Appendix III gives an example of such an interview.

PROBLEMS IN THE LAND-USE SYSTEM

People can almost always name a few key problems they would like to solve. One way to focus a discussion on problems is to start with basic needs and production systems. Needs at the household level include food, water, shelter, fuel, cash income, savings and investment opportunities, assurance of inheritance, raw materials for crafts and resources to meet social obligations. Community members can list problems related to these needs and rank them in order of importance, possibly by a show of hands. However, priorities may differ among different age, gender or economic groups (see Appendix III).

Problems at the community level can be identified by a similar process. These may relate to resource management, transportation, marketing or production. They may be specific to certain places or types of land. After people have identified and ranked their problems, they may have quite a bit to say about what they have done in the past, what succeeded, what failed and why. This information offers valuable lessons for the development of new agroforestry practices.

When a group seems to be ready to move on, or a topic seems to be exhausted, field workers might end the discussion with a few questions. Group members might be asked to rank the most important land-use problems and to think about future forms of land use and resource management, including the uses of trees, choices of tree species and likely tree planting or management sites. People in the group can go home, think over these questions and discuss them with their families and friends. They may agree to come back 1 or 2 weeks later to continue the discussion.

An Adaptive Research Planning Team in Zambia found that people often change their minds about land-use preferences and priorities after talking with specialists or their families, friends or elders. In Siaya District, Kenya, the Cooperative for American Relief Everywhere (CARE)-Kenya agroforestry project staff noted that initial discussions with women tended to focus primarily on exotic species and more commercial or 'official' uses of trees. The same women often came back for follow-up discussions with lists of local tree species for home use. In going about their daily tasks and thinking about all the ways in which they used trees, they completed new, longer lists. The opportunity to think, discuss and come back for further meetings is particularly important in making decisions on the agroforestry products desired—such as fodder, fuel, cash income or fertilizer for crops—, on species preferences and on the availability of various sites for planting trees.

2.5 Household Interviews

A household usually refers to people who share a home, food and wealth: it is not the same as a 'family'. The most practical way to identify households in a community is to start with a house and the people who live in it and to ask if there are others who also belong because they share food, money, labour or decisions on a regular basis. In some cases, field workers may simply ask who eats together regularly, while in other situations several houses with small families may be found together in one home compound and the residents comprise a single household by virtue of shared labour, wealth and decision making.

Interviews with household members in their homes can provide much of the same information as general discussions with community groups. Community discussions give field workers an overview of the different types of household present in the community and the extent to which their land-use practices and interests differ. Household interviews provide more detailed information on specific land-use practices and the cropping calendar (see Appendix III). Such interviews also allow field workers to learn about the division of labour, the sharing of responsibilities, the terms of control and decision making and the access to income and resources within the household. In some cases, it may be necessary to discuss these topics in further interviews with individuals or small groups within the household.

When visiting people's homes, it is usually best to make an appointment in advance, either through local extension agents or informal contacts. However, field workers might also make a few unplanned visits to households which have been suggested by other community members as good examples of particular household types or as especially knowledgeable about specific practices. In any case, it is wise to inform the entire community that these visits will be taking place.

Household members may prefer to seat everyone inside or near the house for the interview, and discussions organized in this way may lead to a lively exchange of information. However, it is often better to move the interview into the fields or to ask to walk around the site after a more formal introductory discussion.

It is usually best if both men and women are present for the first interview. Later, field workers may wish to meet separately with each group to discuss their specific knowledge and skills and their roles as workers, providers and decision makers. The same may be necessary for people in different age groups. This information may be important to develop agroforestry practices that serve all household members.

A general understanding of how a household operates should make it possible to determine who in the household will plant trees, what they will plant and where, who will maintain the plants and who will benefit from the products or improvements provided. It is important to ensure that the work and benefits associated with new agroforestry practices will be fairly distributed without creating or reinforcing any imbalance among members of the household. Small household level discussions, repeated over time, may provide good opportunities for careful examination of how new practices can be conveniently fitted into the time and space available to each person in the household. Discussions on these issues are most productive if repeated and supplemented with individual follow-up interviews on special topics.

2.6 Walking Interviews

In any situation where a variety of people use land and trees in different ways, there is potential for conflict. There is also ample opportunity for efficient complementary systems of resource management for multiple use by a variety of users. It is important to understand the full range of use and user combinations, as well as the potentials for conflict or complementarity, not only to appreciate how conflicts and sharing occur, but also to help plan beneficial changes in the system.

Many questions about who uses land resources and how can be answered by observation and casual interviews while walking through the landscape with individuals or small groups. When passing animals, homes, markets, roads, fences, fields or people engaged in various activities, a field worker can easily ask questions about 'who uses what' or 'who owns what'. Some questions will be listed here: these need not be asked directly but indicate what information is useful and what is reasonable to ask:

- Are the same people using a single place in several ways?
- Are the different uses of the place compatible; are they neutral, mutually reinforcing, competitive or in serious conflict?
- Are different people using the same place, in the same or in different ways?
- Are the people using the same place using the same plant species or different species?
- Are they using the same individual plants or are they each using different individual plants?
- Are the different uses of plants compatible: are they neutral, mutually reinforcing, competitive or in conflict?
- Where there are multiple users, who are they?
- What rights of access and/or ownership do they have in respect to the land, plants or their products?
- Are the users compatible with each other, on good terms, indifferent or in conflict?
- Are they from the same or different households, from the same or different groups, from the same or different places?

The answers to these questions may make the difference between success and failure when introducing agroforestry practices in which land and trees will be shared or used separately by different groups. Walking interviews can also provide an excellent opportunity to discuss changes in the use, management and condition of natural resources and to speculate about the future landscape.

2.7 Individual Interviews

Individual interviews can cover some of the same topics as group discussions and walking interviews. The major difference is that in individual interviews the field worker can arrange to spend more time with one person without interruption and hear the 'whole story' on a particular topic. This is especially important if field workers want a full account of the history of a place or detailed information on a specialized subject, such as herbal medicine, charcoal making, crafts, food processing, marketing, tree nurseries or specific land- or water-management practices.

Discussions with individuals can go into considerable detail on topics of special interest, including explanations that would be awkward or impossible in a group setting or information that a specialist might be unwilling to share in the presence of competing practitioners. For example, women, poor people and members of minority groups may be willing to explain their situation and viewpoint more candidly on an individual basis than would be possible in the context of a group. Many of the same questions can be posed to individuals that were already used in group and walking interviews, although in some cases completely different kinds of questions might be asked. An example is provided in the form of a question-answer sheet in Appendix III, which lists sample questions for a specialized interview with a small group or an individual on the use and management of trees and wild plants.

2.8 Action Interviews (Participant Observation)

Working alongside community members can give field workers an opportunity to see how tasks are performed and to ask questions in a less formal way, as apprentices or helpers. The work performed also makes a positive contribution the individual or group being interviewed, whereas asking people to participate in a long, sit-down interview actually takes time away from their leisure periods or work.. If field workers regularly take part in community work, they may acquire enough skill to make a significant contribution, as well as gaining a place in a household or community work group.

On herding, gathering or water-collection trips, field workers can combine questions about the things they see on the way with discussions of the work itself. These occasions can be especially important sources of information about land ownership, terms of access to shared lands and management of shared resources, as well as providing insight into specific tasks and procedures.

2.9 Timing the Survey Process

The time required for intensive discussions and interviews depends on the requirements of the field workers and the participating community and the complexities of the land-use systems in the area. There is no point in rushing through this important exercise, nor in stretching it out for months while people wait to begin a project. In some cases, it may be possible to spend a full season on initial discussions, while in others field workers may need to begin new activities more quickly or continue with existing work. After a month of mapping, field visits and group interviews, enough information should be available and enough interest generated to start planning which trees to grow and where and how to grow them. In the meantime, field workers can continue to gather more detailed information through occasional individual interviews and work sessions with specialists or particular groups.

Field workers must decide what information is needed on a priority basis in order to use the first intensive period of interviews and field visits most effectively. Although a minimum of information can be obtained in a few weeks, the learning process and partnership with the community should continue throughout the life of the programme.

What matters is that field workers meet two major objectives before going on to plan and implement new agroforestry practices. The first is to establish a partnership with the local people based on respect for their knowledge, experience and priorities for the future. The second is to collect practical information which community members can use in selecting appropriate agroforestry practices.

2.10 Using the Survey Results to Select Agroforestry Practices

After field workers have gathered information on different topics through the survey activities described here, they need to review and summarize this information, drawing also on their own previous knowledge and training. They should then prepare their information in an appropriate form for review, discussion and revision by the community. After community members have had time to consider and discuss alternatives, they can choose the most suitable species and agroforestry practices for local conditions, priorities and resources and begin work plans for the next planting season.

Although summarizing all the information gathered during the surveys may be difficult, it is possible to make a simple, clear, general summary for presentation to the whole community. The form in Appendix IV may be used to list the most important types of land and land uses, the major groups of land users, the highest-priority uses of trees, shrubs and grasses and the most serious problems at the household and community levels. Once this information is listed, field workers can review the special constraints and opportunities related to tree planting which people mentioned during the survey.

At this point, field workers have to make judgements, based on information collected during the survey, about potential conflicts in the use of land and trees and about the most promising species and most appropriate sites for new agroforestry practices. What are the obstacles to agroforestry practices in this community? Are there any sites or types of site where serious conflict might arise over the use of land and trees? After considering these questions, field workers should list the most popular local species which will serve the uses identified as priorities (see Appendix IV). This summarized information can be presented to representative community groups to review and revise before presentation to the community as a whole.

For more detailed planning and decision making, it is usually necessary to divide the community into separate land-user groups, based on each group's specific needs and resources in terms of agroforestry practices. The result should be a list of all the 'situations' which require a separate approach. For example, a community might have five different groups for the purpose of agroforestry: farmers with access to permanent water and fertile land, farmers with dry, poor-quality land, landless women farm labourers, resident herders and nomadic herders who use some of the land on a seasonal basis.

Field workers can make short summaries for each special group and situation similar to the general summary made for the community (see the form in Appendix IV). Representatives of each group should review the relevant summary and add their comments and revisions. Once all of these descriptions are complete and confirmed, they can be distributed in writing to teachers, leaders, and interested members of the community. However, if potential conflicts between different groups are too sensitive, then field workers may choose to distribute only the general summary for all the groups combined.

Once there is some agreement about the basic conditions for the introduction of agroforestry practices, then it is easier to use the detailed information available on local species, including their uses and potential for different kinds of agroforestry systems. The original interview notes will include descriptions of priority uses, key problems and major obstacles to agroforestry practices (see Appendix IV) which field workers can compare with the information matrices they compiled on local species. A list of the trees, shrubs and grasses which best meet the most important needs and conditions can then be discussed once again with community groups.

Field workers can also check the supplementary lists of species in Appendix I in order to identify any other species, new to the community, that could provide products or services not already available. They may identify some species which grow well under difficult conditions or which are particularly compatible with crops, pastures or other trees. Each of these lists should be carefully revised to be sure that the species are appropriate for the environmental and land-use conditions of the community group for which the agroforestry practice is being planned.

The next step is to check the list of agroforestry practices in Appendix I against the species lists and the list of planting spaces available for each community group. To determine the practices most suitable for local conditions, field workers should review the appropriate sections of Chapter 3. Members of each community group need to discuss their own suggestions for species and land-use practices: field workers should try to encourage lively discussion and debate, making sure that everyone gets a chance to contribute.

Once group members have made their suggestions, field workers may want to add their own ideas on appropriate agroforestry species and practices. These should be presented as contributions to the discussion, not as conclusions. In the end, the community groups must review all the suggested practices and species and discard those options which do not stand up to closer scrutiny. People will probably think about their choices and discuss them at home before deciding, possibly by vote at a later meeting, which practices and species to try out during the next planting season.

If field workers are planning a formal research project, it may be important to identify a small number of agroforestry products, services and species for evaluation. However, informal exploration and community-based research allow for much greater variation. People may want to try 'a little of this and a little of that' until they find species and practices they wish to test in more detail. In many cases, they may decide to try out a large number of species for several different uses and may fit these species into three or more practices at different types of site.

Once these decisions are made, participating community groups and field workers will have a list of which practices and species to try first. The information provided in Part 2 and Appendices I and II will now be more useful, along with all the information collected during the survey process. Field workers should continue to meet with people as they try out agroforestry practices and should encourage them to make changes and improvements to suit their own needs and preferences. Chapter 3 explains the process of evaluation and how to make it part of the entire cycle of learning, decision making and testing.*

---

*The material in this chapter was excerpted from Rocheleau, D. (1988).

SELECTED REFERENCES

Atta-Krah, A.N. and Francis, P.A. (1987). The role of on-farm trials in the evaluation of composite technologies: the case of alley farming in southern Nigeria. *Agroforestry Systems.* 23: 133-152.

Buck, L. (1988). *Agroforestry extension training sourcebook.* New York: CARE International, 540pp.

Chambers, R. (1981). Rapid rural appraisal: rationale and repertoire. *Public Administration and Development.* 1: 95-106.

Chambers, R. (1983). *Rural development: putting the last first.* Harlow, UK: Longman, 246pp.

Chambers, R. and Jiggins, J. (1987a). Agricultural research for resource-poor farmers. Part I: transfer-of-technology and farming systems research. *Agricultural Administration and Extension.* 27: 35-52.

Chambers, R. and Jiggins, J. (1987b). Agricultural research for resource-poor farmers. Part II: a parsimonious paradigm. *Agricultural Administration and Extension.* 27: 109-28.

Conway, G. and McKracken, J. (1988). *Agroecosystems analysis.* London: IIED, 35pp.

Farrington, J., ed. (1988). *Farmer participatory research.* Reprinted from *Experimental Agriculture*, 24: 269-342. London: Overseas Development Institute, 73pp.

Feldstein, H., Buck, L. and Rocheleau, D. (in press). The Siaya agroforestry project in Kenya. In H. Feldstein, ed. *Intra-household dynamics and farming systems research and extension: a casebook.* Hartford, Connecticut, USA: Kumarian Press, 165pp.

ICRAF (1983a). *Guidelines for agroforestry diagnosis and design.* Working Paper No. 6. Nairobi: ICRAF, 25pp.

ICRAF (1983b). *Resources for agroforestry diagnosis and design.* Working Paper No. 7. Nairobi: ICRAF, 383pp.

Jiggins, J. and Feldstein, H. (in press). *Methodologies casebook on intra-household issues in farming systems research.* Hartford, Connecticut, USA: Kumarian Press, 180pp.

Raintree, J., ed. (1987). *D&D user's manual: an introduction to agroforestry diagnosis and design.* Nairobi: ICRAF, 110pp.

Rocheleau, D. (1988). *A community-based approach to agroforestry research and development.* Nairobi: Ford Foundation, 120pp.

Rocheleau, D.E. and Malaret, L. (1988). Use of ethnoecology in agroforestry systems research: an example of agroforestry technology and pest management research in Kenya. In W. Miller, ed. *Proceedings of the annual farming systems research symposium.* Morrillton, Arkansas, USA: Winrock International, 300pp.

Rocheleau, D.E., Wachira, K., Malaret, L. and Wanjohi, B.M. (in press). Ethnoecological methods to complement traditional knowledge and farmer innovation in agroforestry. In A. Pacey, R. Chambers and L. Thrupp, eds. *Farmer innovation and agricultural research.* Brighton, UK: University of Sussex, Institute of Development Studies, 280pp.

Scoones, I. (1988). *Community management of indigenous woodland project: Zvishavane and Chivi Districts, Zimbabwe.* Harare: ENDA, 115pp.

Sundberg, J. and Okali, C. (1988). Farmers, on-farm research and the development of new technology. *Experimental Agriculture.* 24: 333-342.

Wilson, K. (1987). *Community management of indigenous woodland project.* Harare: ENDA, 115pp.

CHAPTER THREE

EVALUATING THE PROJECT

Research and development agencies often reserve evaluation for the last stage of a project or programme, as a kind of final judgement on its success or failure. It is common to hire external evaluators to make these assessments. They usually measure the success of rural development projects by the quantity and quality of products obtained, by the cash earnings realized by the participants or by the area of land affected, such as number of hectares reforested or planted to a new crop.

However, the evaluation of an agroforestry project does not have to be limited to a seal of approval or a critical judgement from outsiders. Field workers, planners and participating communities can evaluate the effects of new agroforestry practices as part of a continuous process. This kind of participatory evaluation can help direct the exploration and application of agroforestry and related activities, to keep real practice in line with needs and priorities.

As the people in a community experiment and learn, regular evaluation sessions provide the means to apply new experience and ideas to the next season's activities. This might mean introducing a new practice, replacing a practice with something better or modifying existing practices. Aside from the agroforestry practices themselves, research and development workers and community members can review all their activities and take steps to make their work more productive, to reach more people or to ensure a fair distribution of benefits. Discussions might cover any aspect of the innovation and development process, for instance the organization of meetings, the training of special technical assistants for women's groups or the production of instructional materials in the local language.

Evaluation, like the planning process described in Chapter 2, depends on asking the right questions, getting honest answers, summarizing the results and applying the information to future activities. Questions should be posed to field workers, to key individuals in the community and to groups that represent different interests within the community. Both agroforestry workers and community members will review and discuss their own work and the way in which they have worked together. In addition, field workers will identify agroforestry practices and community activities that might usefully be introduced in other places.

As in the initial interviews and field visits, evaluation begins and ends with questions. The first questions explore the situation and the final questions concern specific practices, followed by practical efforts to test new practices and new ways of working. So evaluation does not end the journey, it just helps to chart a new course every so often as conditions change.

Agroforestry activities may be evaluated once a month, once a season or once a year. In fact, almost every meeting or visit involving groups of participants can include some evaluation, with more focussed

discussions at specially scheduled evaluation meetings. Of course, some issues cannot be evaluated until trees are mature, which could be several years after planting.

Several types of evaluation activities might be appropriate. These could range from visits to farmers' fields to see how they have carried out or modified different agroforestry practices to group meetings and household or individual interviews. Periodically, it might also be useful to look out over the landscape and chart the distribution of participants or the areas where a particular practice has succeeded or failed.

Field workers and community members should ask questions covering both agroforestry practices and their own activities, referring to the five main points of the land-user perspective outlined in Chapter 1: How well do agroforestry practices and activities address multiple uses, multiple users, the changing landscape, local knowledge and experimentation and land users as clients and co-workers? Questions should be wide ranging, rather than confined to a particular list, but the following questions suggest some of the information that needs to be obtained in order to evaluate an agroforestry practice.

How well does the specific practice, or mix of practices, deal with multiple uses? Does it produce the right products and services? Check this against the original list (Appendix IV) from early discussions with the community. Does it produce enough? Is the quality good? How good is the timing of production or services? Is the amount and quality of products or services worth the work, land and other investments required? Have other products or services become important that should now be provided by this or some other practice?

Has this agroforestry practice, or combination of practices, addressed all of the land-user groups in the community? If not, are there complementary practices to meet the needs of other groups? Has this practice harmed one group? Has it created any new conflicts or resolved any pre-existing ones? Has it changed the relationship between different groups, for instance by sex, age or economic level? Has anyone gained or lost rights of access, use, harvest or ownership? If so, who? Did the same people who did the work reap the rewards? Did any group work more or benefit more than the others? Are there suggestions about how to resolve any difficulties or problems encountered?

Has this agroforestry practice, or combination of practices, offered an adequate solution to the environmental problems identified in earlier surveys and discussions? If not, what other practices might be introduced? Are the agroforestry systems in the right locations? Do the shapes and arrangements of plants fit peoples' preferences for the present and future landscape? Are there any spaces or types of placement that have caused problems? Have any places been redefined because of agroforestry practices, such as shifts from common to private land, from open to restricted grazing or from men's to women's domain? Are there any new ideas or local experiments concerning design, the use of space or the combination and placement of trees? Are there any major changes in the landscape or new trends that might affect the design and placement of agroforestry systems?

Did the agroforestry practice take into account the community's prior knowledge and experience? Has it led to any improvements? Were local ideas and experiments related to this practice correctly understood and interpreted? Was the community's knowledge and practice documented and returned to them in an accessible form for reference or educational use? Have people in the community received enough new information, in the right form, to understand and evaluate the new practices? Would they like more or better information on any points or topics? Have community members or groups of specialists incorporated any of the new information or innovations into their own practices? For example, have herbalists integrated new species into their remedies or have people incorporated new species and practices into their traditional classifications of plants and land use?

Another set of questions focusses on how field workers and community members have worked together. Did field workers consult the community adequately and did they interpret the answers they received about local needs and priorities correctly? Did they involve people as partners in the selection and testing of agroforestry and related practices? Are there ways that they might do this more effectively? Did research and development workers give back as much as they took in terms of time and information? Are community members better prepared to conduct their own experiments and extension programmes than they were before? Are their knowledge and experience well documented for future use? Do community members have better links with each other and with outside resources than they did before?

Have local people responded to initiatives by field workers or members of their community to collect information, discuss problems, decide upon solutions and test agroforestry practices? Have they voiced complaints or suggestions for improvements? Have they contributed new ideas on agroforestry or land management for testing by the community? Have they cooperated with each other in interpreting the results of trials? Have they shared any of their experiences with friends, relatives or officials in neighbouring communities?

As community members and field workers answer these questions, they need to decide whether the results mean that a particular practice should be continued, changed, supplemented by another practice or dropped. Are the answers widely applicable or do they relate only to specific people, land-use systems or environmental conditions? For each situation defined in the original surveys, a summary should be made of each practice introduced, including the conclusions reached during the evaluation. These conclusions should be discussed and revised by all participants and used to formulate work plans for the following season.

At some point, field workers might also visit a neighbouring community or another area where the agroforestry practices developed during the project might be adapted and introduced. Eventually, this process can be taken over by the extension service or by formal or informal networks of community leaders, educators and other development workers. When the people in the area no longer need specialized agroforestry services from outside, but are prepared to continue and expand the work they began together, then a community-based agroforestry development effort may be considered a success.

SELECTED REFERENCES

Buck, L., (1988). *Agroforestry extension training source book.* New York: CARE International, 540pp.

Buck, L. and Ngugi, A., eds. (1988). *Agroforestry monitoring and evaluation project: regional workshop report.* New York: CARE International, 235pp.

Feuerstein, M. (1986). *Partners in evaluation: evaluating development and community programmes with participants.* London: MacMillan, 240pp.

Rocheleau, D. (1985). *Criteria for re-appraisal and re-design.* Working Paper No. 37. Nairobi: ICRAF, 24pp.

Scherr, S. and Mueller, E. (1988). *Technology monitoring and evaluation in agroforestry projects: a review.* Nairobi: ICRAF, 40pp.

PART TWO

AGROFORESTRY PRACTICES

This section describes 15 agroforestry practices. Several of these are likely to be appropriate for any local need or environment and they can be modified to suit a particular situation. Some practices are similar to each other and at times the distinction between them may seem arbitrary. In fact, these agroforestry practices form a continuum—as the landscape does—and can fit together in various ways, depending upon the environment and the goals of the local community.

Each chapter in this section begins with a general description of a particular agroforestry practice—what it looks like and what it is used for. Next, design factors are discussed, including location, layout, appropriate tree and shrub species and management techniques. This is followed by a summary of potential benefits and examples from different locations in dryland Africa. Finally, there is a list of selected references. The information presented in these chapters is, where possible, based on actual experience in the subhumid and semiarid zones of Africa.

Throughout these chapters, social considerations are pointed out as well as environmental constraints. Proper management practices—from care of seedlings in nurseries to pruning when trees are mature—are important to all the agroforestry practices described in this book, but these are not specified in any detail. Other manuals covering these topics are usually available through government forestry or agricultural extension services or development agencies. However, water management for agroforestry practices is particularly important in dryland areas and some suggestions have been made in Chapter 5.

Tables 1 and 2 in Appendix I provide suggestions for tree and shrub species suitable for different situations. These species are also listed in Appendix II. However, this is a short list compared with the hundreds of tree and shrub species suitable for use in dryland Africa: The most important species are those already growing and in use in a local area. They can be listed in the format suggested in Table 3. These trees and shrubs have been tried and proven under local conditions and are likely to form the basis for any agroforestry system designed for local use.

CHAPTER FOUR

AGROFORESTRY IN CROPLAND

4.1 Dispersed Trees on Cropland

DESCRIPTION

Trees may be grown on farmers' fields while crops are grown in the understorey. The trees may be dispersed widely, either spaced systematically in a grid or scattered at random. This practice is distinct both in form and in purpose from agroforestry based on trees and shrubs planted in lines, as described in the sections on contour vegetation strips (section 4.2) and on alley cropping (section 4.3).

The practice of raising trees dispersed on cropland may be based on protection and management of selected mature trees already on the site, it may involve planting new trees or it may depend upon careful

Acacia albida *over sorghum and maize, with pod baskets.*

management of selected seedlings established on site through natural regeneration. In the Sahel, species commonly used for this purpose are *Acacia albida, Butyrospermum parkii* (karite), *Parkia clappertonia, Parkia biglobosa* (nere), *Borassus aethiopum* (palm) and *Acacia senegal.* In Eastern and Southern Africa, farmers use *Markhamia platycalyx* (Siaya District, Kenya), *Acacia albida* (Southern Province, Zambia), *Cordia abyssinica* (East African highlands), *Sesbania sesban* (Kakamega District, Kenya) and *Croton macrostachys* (Central Province, Kenya).

In these different situations, the function of the trees may vary. Usually they provide a product of commercial or subsistence value, such as food, fuel, oil, building poles, fodder or gum. The nature of the product or of the trees themselves often requires that they be maintained individually rather than managed in a hedgerow or a crowded lot. In some cases the trees also provide a *service*—they improve soil fertility, conserve soil moisture or otherwise improve the microclimate, resulting in increased crop yields.

DESIGN

Farmers may have several reasons for growing trees dispersed in cropland. Some of the most common are:

- to increase crop production
- to extend the time that a particular field can be cropped
- to increase the total yield of mixed products from the cropland
- to diversify the range of products
- to produce a particularly valuable product in a secure site, protected from animals.

Often farmers combine two or more of these objectives. For example, they may keep *Acacia albida* in their fields partly as a source of dry-season fodder which can be sold or traded to herders. After the crops are harvested, herders are allowed to bring in their livestock to congregate and deposit manure on the fields.

Spacing is determined by the size and requirements of the trees and also in order to fit the trees into the cropland in a way that minimizes interference with crop cultivation and that makes the best use of any positive effects of the trees on crops. The choice of tree species and pattern and density of placement varies according to individual circumstances, but some general guidelines can be drawn from well known examples. In maize- and millet-based cropping systems in the Sahel, tree density ranges from 40 to 60 trees per hectare for *Acacia albida*, 60 to 80 per hectare for *Parkia clappertonia, P. biglobosa* and *Butyrospermum parkii*, 200 per hectare for *Borassus aethiopum* and up to 300 per hectare for *Acacia senegal.* New stock is normally planted with the following spacing: *Acacia senegal* at 4 x 4 metres, *Borassus aethiopum* at 7 x 7 metres, *Parkia biglobosa* and *Butyrospermum parkii* from 7 x 7 to 10 x 10 metres and *Acacia albida* at 10 x 10 metres.

Similar spacing of dispersed trees can be found in more humid environments where trees are used as shade and protective cover for crops. On cash-crop plantations, spacing may be systematic (for example, 7 x 7 metres for coffee shade), while in other situations, spacing may be less formal.

Not only do density and spacing vary by tree species, but also according to the surrounding cropping system. Density reflects the relative value of the tree products versus the crops, both at home and in the market, while spacing is determined in part by the positive or negative effects of the trees on crop yields. If any form of mechanization is foreseen during the lifetime of the trees, which varies from 20 years for *Acacia senegal* to over 100 years for *Acacia albida*, they must be planted in reasonably straight lines with ample room for equipment to move between them.

Farmers establishing seedlings in a closed field.

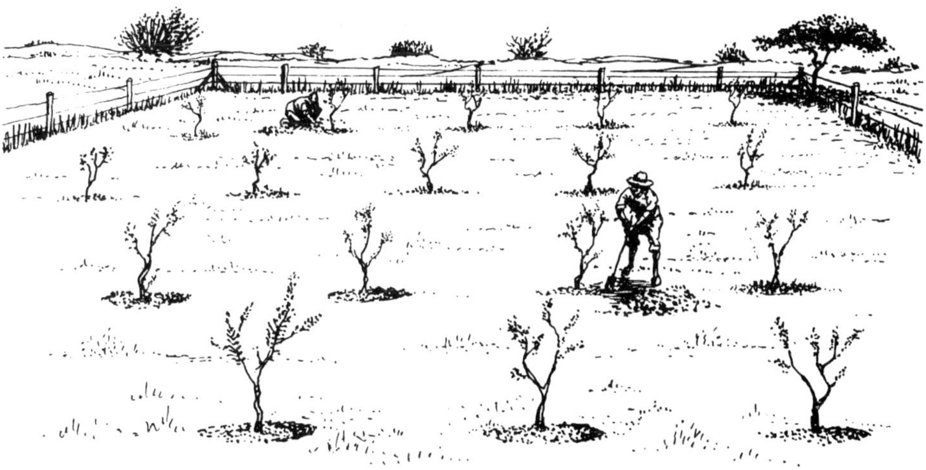

Trees growing in cropland should compete only minimally or not at all with crops and, if possible, should contribute to crop growth. Tables 1 and 2 in Appendix I indicate species with desirable characteristics. These include a deep rooting system, a form that produces only light shade, a capacity to improve the soil through nitrogen fixation and/or leaf litter and no tendency to harbour crop pests. Farmers also keep trees on cropland to obtain tree products, such as fruit or fuelwood. However, trees may harbour birds or other pests which can damage field crops.

Several species have been suggested for planting in cropland: the most well known is *Acacia albida.* There are many others, both exotic and indigenous. West African farmers often use *Terminalia* spp. as shade for coffee; *Cordia abyssinica* and *Grevillea robusta* are among the species used in East Africa for the same purpose. Some authorities discourage the use of *Grevillea robusta* over coffee and tea, as they suspect that this species may harbour crop pests and diseases which affect these crops.

Additional tree species are becoming available with greater productivity and more direct and substantial benefits in terms of soil fertility. In almost all situations where trees are grown dispersed in cropland, the main conservation goal is to provide more organic matter and/or nitrogen to the soil and to improve the microclimate for crops.

ESTABLISHMENT

In the Sahel and East Africa, trees in cropland originate mainly from natural regeneration or from seedlings. Most of the commercial species used, with the exception of *Borassus* palms, are now propagated as seedlings in nurseries. Over the past 20 years, direct seeding has also been used in many situations, sometimes with notable success. In one situation in West Africa, at a government forest reserve near Matameye, Niger, the local forest agent fed *Acacia albida* pods to a flock of sheep which he then pastured on the area where he wanted regeneration to occur. This experiment was undertaken about 25 years ago and the standing trees are the visible result today.

Borassus is 'seeded' by slightly burying the entire fruit and allowing it to germinate at the site. Experiments in Senegal, Chad and Sudan have indicated that *Acacia senegal* seeds can be treated and sown in hills or by broadcasting if rainfall is sufficient. The seeds are sown in seedbed strips approximately 1 metre wide, with the soil surface lightly scarified or dug to remove some of the existing ground cover and to increase water infiltration.

Farmers in Kenya's Nyanza Province plant *Markhamia platycalyx* in blocks near the home compound. The trees mature in about 15 years and are harvested for building poles. Crops are then planted between the coppicing trees which produce small poles, wood and fodder.

Where natural regeneration occurs, it may not be necessary to plant seedlings. Rather, the young trees which regenerate may need to be protected from grazing animals, fires or land-clearing activities. Several successful projects have been based on better protection of natural regeneration, especially for *Acacia albida* and *A. senegal.* Specialists who have worked with these trees for years have concluded that it is better to protect young natural stands than to plant nursery stock.

The practice of protecting existing, natural regeneration rather than raising trees in nurseries and then transplanting them, has many advantages. However, one disadvantage is that trees will not be properly aligned for crop cultivation by mechanization or animal traction. Trees already in place can usually be thinned to allow for mechanization.

MANAGEMENT

During the first year of growth, trees in cropland need to be protected from animals, either individually or by protecting the entire field. The area surrounding young trees also needs to be weeded and may require other improvements, such as the construction of microcatchments in dry areas (see section 5.4).

Mulching with leaf litter may help young trees, but in many areas of West Africa mulch attracts termites. However, some mulches, such as wood ash, *Azadirachta indica* leaves or seed cake, may kill or repel insects as well as providing soil cover and enrichment.

Three of the many ways seedlings can be protected.

ANTICIPATED BENEFITS

Farmers in the Sahelian region of Senegal reported that yields improved when crops were planted with *Acacia senegal, A. albida* and *Borassus aethiopum*, especially directly underneath or near the trees. Some more formal studies (notably the work done by Charreau and Vidal in Senegal) have confirmed this: in a controlled setting, sorghum crop yields were 60% higher in fields with *Acacia albida.*

Trees planted in cropland also provide several products of value to farmers. *Acacia albida* provides pods for supplementing livestock feed during the dry season when forage is scarce, branches for thorn fencing and wood for utensils, such as bowls, water troughs and large mortars used for cereal pounding in many parts of the Sahel. *A. albida* wood is also used in Northern Kenya for bowls and water troughs. This species has the added advantage of shedding its leaves during the crop growing season. Thus it does not shade the growing crops, yet provides shade to the soil during the dry season.

In many areas, the *Borassus* palm furnishes the only available construction material that resists termites and rot, used to build the flat roofs traditional in much of the Sahel. *Borassus* also provides food: Not only are different portions of the fruit consumed at different stages of ripening, but the young shoots, which grow underground, are considered a local delicacy and can be sold at a good price. The fronds are used for woven mats, roofing, fencing panels and walls for houses, while *Borassus* frond stems make excellent fencing material because they are sturdy, long lasting and thorny.

Farmers report that these palms do not interfere with crops growing underneath during the first 5 to 10 years of growth. If the crowns grow too large and shade the crops, a few fronds can be cut from each tree and used for weaving. Over the next 10 to 15 years, when cropping is impossible directly underneath the trees, a grass cover becomes established that makes excellent pasture for animals. After the trees have grown taller and certainly once they are 40 years old or older farming underneath them can take place completely unhindered and without any loss in yields. In fact, crops planted near *Borassus* palms seem to give higher yields even in areas where tree densities are in the order of 300 to 400 stems per hectare, such as *Borassus* stands in the Bana forest, near Gaya, Southern Niger.

Acacia senegal provides gum, fuelwood and fodder. The fodder is especially important in years when grass cover is sparse because of poor rainfall. *A. senegal* also fixes nitrogen, improves the condition of the soil and its wood is regarded as one of the best available in dry areas.

Parkia biglobosa and *Butyrospermum parkii* both produce fruits that are used extensively as a staple food in substantial areas of Senegal, Cameroon, Chad, Guinea, Ghana, Ivory Coast, Togo, Benin and Nigeria. *Butyrospermum parkii* produces a butter that is appreciated as far away as Japan, while the pods from the *Parkia biglobosa* tree are used in a sauce that accompanies the staple cereal dish in extensive rural areas.

Markhamia platycalyx provides high-quality wood for construction and furniture, as well as a limited quantity of fuelwood as a by-product. Its leaves are used as fodder in times of drought. Many farmers in Siaya District, Kenya, are willing to give up some cropland space to grow these trees on the basis of the economic returns from their products.

Some evidence also suggests that this species may have a beneficial effect on soil fertility.

These six tree species are a mainstay of life in the Sahel and other dry areas of Africa. Aside from these well-known species, there are scores of other trees occupying similar niches in site-specific land-use systems throughout the continent. These local trees may be better suited to people's needs and conditions than any of the exotic species mentioned in this manual. It might be worthwhile to compare different species, using a mix of widely used trees with local ones. Chapter 2 and Appendix III provide some suggestions on how to identify likely candidates for trials.

COMBINATION WITH OTHER TECHNIQUES

Windbreaks may be used with dispersed trees in cropland, especially using *Acacia albida* (see section 6.3). Dispersed trees may also be combined with earthwork structures (section 5.1) and microcatchments (section 5.3), especially for soil-moisture conservation in dry areas.

EXAMPLES FROM THE FIELD

The maintenance of dispersed trees in cropland is a traditional practice throughout Africa. One of the first development efforts which made use of this practice was the reintroduction of *Acacia albida* in

A Luo homestead with a Markhamia *stand that has been coppiced.*

1930 in groundnut-growing areas of West Africa. Since then, similar efforts have been undertaken throughout the Sahel, most notably by national governments, voluntary organizations and international agencies, such as the United Nations Development Programme (UNDP), the Food and Agroculture Organization of the United Nations (FAO) and the United Nations Educational, Scientific, Cultural Organization (UNESCO), in Niger and Chad.

As a result of these activities, the negative reputation of *Acacia albida* as slow growing has been modified. *A. albida* planted near Madarounfa, Niger, in 1972 grew 5 to 7 metres tall in 11 years, with trunk diameters at breast height ranging from 15 to 25 cm. Pod production was already substantial during this period and crowns were beginning to spread, covering an average of 4 to 6 square metres. Several *A. albida* plantations were initiated from 1974 to 1977 through a project sponsored by CARE between Ndjamena and Bougor in Chad. These trees are in excellent condition today, with growth similar to that reported in Niger.

Acacia senegal has been reintroduced in Senegal, Chad and Sudan, where its value has been demonstrated. In northern Senegal, the local pastoral people are now planting and maintaining these trees without any outside project support.

Parkia biglobosa and *Butyrospermum parkii* have also been reintroduced successfully in many locations. At some sites, for instance around Ouahigouya and Niamey, Niger, *P. biglobosa* had disappeared from the landscape in the past 40 to 60 years. Some of the reintroduced trees have been put under considerable stress by people removing bark from the trunks to produce a traditional medicine, for example in the 'green belt' around Niamey planted on a trial basis in 1965.

The Centre Technique Forestier Tropical (CTFT) in Nogent-sur-Marne, France, and other organizations have carried out studies on *Borassus* palms, and stands of these trees have now been reestablished in southern Niger and in Senegal. Other work with this species has focussed on improved management and the encouragement of natural reproduction.

Markhamia platycalyx is less widespread and its use not as well documented as that of the other species mentioned. CARE-Kenya, in collaboration with women's groups and other self-help organizations, is encouraging the practice of cropping between stands of this species and planting new trees in cropped fields.

CONSTRAINTS

The major constraint on the propagation and management of trees in cropland is pressure from grazing animals. Seedlings of many species have a degree of tolerance for browsing and fires, but when grazing

Borassus palm plantation with grazing cattle.

pressure becomes too intense natural reproduction no longer occurs. In many areas of dryland Africa, there simply are no more young trees.

Mature trees are killed by drought. For example, large-scale losses of *Acacia senegal* resulted from the droughts of the early 1970s. Accounts from reliable sources suggest that up to 80% of all *A. senegal* in existence before this period have died and that no substantial natural regeneration has occurred in areas subjected to prolonged drought.

SELECTED REFERENCES

Ahmed, A. (1979). Effects of land use on soil characteristics in the Sudan. In H.O. Mongi and P.A. Huxley, eds. *Soil research in agroforestry* Nairobi: ICRAF, pp. 1-13

Anger, G., ed. (1985). *L'arbre et la haie dans l'exploitation paysanne.* Fiche Technique No. 3. Nyabisindu, Rwanda: Project Agropastoral de Nyabisindu, 126 pp.

Depommier, D. (1985). The role of the woody vegetation in soil conservation and rehabilitation. In *Proceedings of a workshop on soil conservation on grazing lands.* Nairobi: Ministry of Agriculture and Livestock Development, Soil and Water Conservation Division, pp. 82-118.

Felker, P. (1981). Uses of tree legumes in sem-iarid areas. *Economic Botany* 35 (2): 174-86.

Le Houerou, H. (1980). Agroforestry techniques for the conservation and improvement of soil fertility in arid and semi-arid zones. In H. Le Houerou, ed. *Browse in Africa.* Addis Ababa: ILCA, pp. 433-35.

Miehe, S. (1986). *Acacia albida* and other multipurpose trees on the Fur farmlands in the Jebel Marra Highlands, Western Darfur, Sudan. *AgroforestrySystems.* 4: 89-119.

Poulsen, G. (1981). *Malawi: the function of trees in small-farmer production systems.* Rome: FAO, 65 pp.

Teel, W. (1985). *A pocket directory of trees and seeds in Kenya.* Nairobi: KENGO, 151 pp.

von Maydell, H.J. (1983). *Arbes et arbustes du Sahel: leurs caractéristiques et leurs utilisations.* Eschborn: GTZ, 530 pp.

Weber, F. with Storey, C. (1986). *Reforestation in arid lands.* Arlington, Virginia, USA: VITA, 335 pp.

4.2 Contour Vegetation Strips

DESCRIPTION

In many traditional African farming systems, living barriers of grasses, lines of stone or wood or 'trashlines' of sticks, leaves and other organic debris are placed across hillsides to control storm runoff and soil erosion. Combinations of trees, shrubs, grasses and creeping vines planted on the contour can serve the same purpose, resulting in greater structural stability and at the same time providing a higher yield and diversity of useful products.

The contour strip is also known as a barrier strip or hedge, horizontal vegetation strip, contour hedge or horizontal hedgerow. It is an erosion-control measure for sloping farmland, which, in addition, provides useful products and enriches the soil. Although contour vegetation strips may be confused with alley cropping (see section 4.3), the two are quite distinct: Alley cropping focusses on improving soil fertility and crop microclimate rather than on preventing erosion. Contour strips also differ from living fences (section 6.1) and boundary plantings (section 6.2), which are generally grown on property lines, between fields or around compounds and are intended primarily for animal control or boundary marking.

The establishment and maintenance of horizontal strips of vegetation on sloping ground is one of the most direct, cost-effective and ecologically sound erosion control-interventions. Controlling erosion by completely covering bare slopes with permanent vegetation would prevent crop cultivation, while building terraces or other structures is labour intensive and can be expensive (see section 5.1). An acceptable compromise is to establish a series of permanent vegetation strips on the contour within a cropped field.

Contour vegetation strips may be planted or they may consist of natural vegetation left to grow across the slope. If the strips are dense and wide enough, they can stop water from flowing downhill and trap soil particles in a web of vegetation and litter. If the soil is permeable, water can then soak slowly into the ground.

Hill-slope farmers, who are aware of erosion hazards and wish to keep more water and topsoil on their fields without resorting to expensive structures, often plant strips of vegetation on the contour. Farmers tend to avoid building soil-conservation structures unless necessary, for instance on very steep slopes, since the labour and materials required are usually expensive and scarce or even unavailable. Most physical structures also reduce the total area available for plant production, although this situation can be improved by adding agroforestry practices (see section 5.1). By contrast, vegetation barriers substitute different productive plants, chosen to protect and stabilize the site, for the crops planted in between.

Acacia albida *in a wide vegetation strip.*

While strips of woody plants are a fairly permanent feature, they are usually more compatible than are earth or stone structures with fallow or rotational cropping systems. Depending on the species used, contour strips can withstand limited grazing during a rotation better than terraces, ditches or contour bunds. The plants on the strip may also yield fuelwood, timber, fruit and other products during the rotation.

Contour vegetation strips are particularly useful in areas of dense population or limited access to off-farm resources. Here, livestock fodder and other products derived from perennial plants may be in short supply and thus highly valued. Contour grass strips in cropland are used from the Kenyan midlands and highlands to the hills of Southwestern Zaire and Rwanda. Contour strips with woody plants, grasses and creepers have been incorporated into a number of traditional systems, for instance in Senegal and on the central plateau of Tanzania.

DESIGN

Vegetation strips on the contour are not always adequate as a hillslope soil-conservation measure and they must be designed carefully. For example, if individual trees are simply planted on the contour, water running between the stems may erode the soil and create rills and gullies. Closely spaced single lines of *Leucaena* have been planted on contours in Malawi, but the effect on adjacent crops has yet to be determined. Depending on site conditions and design, strips may be used alone or with supplementary earthwork structures and drainage channels. All the erosion-control measures discussed in this book should be considered when designing an on-farm erosion control system.

Woman gathering 'cut and carry' fodder from a grass strip.

Erosion between tree stems planted on the contour without soil conservation structures or adequate ground cover.

Aside from their design, the effectiveness of contour vegetation strips depends in large part on slope, rainfall intensity and soil conditions. For conserving soil and water, the two most important factors are the width of the strips and the intervals between them. In general, steeper slopes require narrow strips spaced close together. On very steep slopes, the strips would be too close together to allow cropping unless they are combined with physical structures. Likewise, contour strips are more effective on deep, permeable soils than on heavy clays.

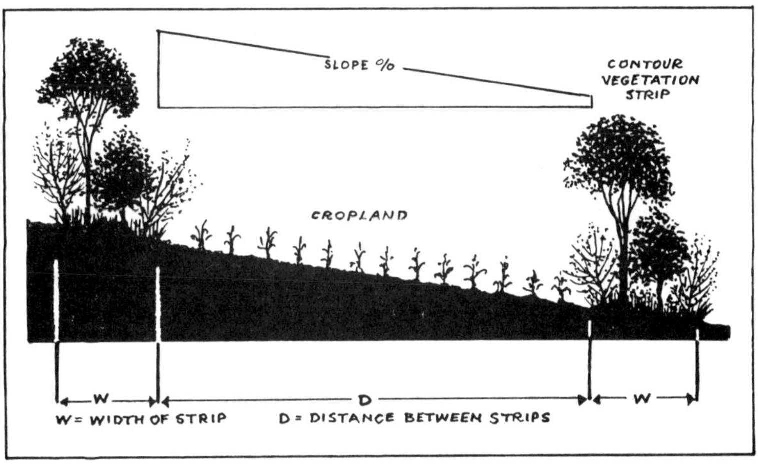

Where soils are less permeable, there is a greater need for barrier and infiltration structures to intercept water runoff.

Contour strips usually range from 3 to 8 metres wide and cover about 10% of the total area under cultivation. They should be as continuous as possible. Properly spaced, they can define the outline of gradually evolving terraces or small earthwork structures (see sections 5.1 and 5.2). In such cases, the strips consist of one or two rows of shrubs and trees with at least one line of tightly spaced grasses planted on the downslope side to trap eroding soil. If no specific design methods are available, the table given here can be used to estimate proper spacing.

Slope (%)	Width of Strip (m)	Distance between Strips (m) Erodible Soils	Cohesive Soils
5	1.5	45	80
10	2.0	35	70
15	3.0	30	60
20	4.0	26	53
30	6.0	23	44
40	8.0	20	36
50	10.0	17	30
60	12.0	14	26
80	16.0	13	22
100	20.0	10	20

SPECIES

Tree and shrub species used on contour strips must be compatible with surrounding crops and cultivation practices. Invasive, aggressive or pioneer species should be avoided. If crops are planted that require a lot of light and the distance between strips is narrow in order to provide the necessary erosion protection on steeper slopes, then species must be selected that do not provide shade. In such cases, hedges that provide fodder or fuelwood may be preferable to larger trees that provide fruit or building poles.

In the savannahs of East Africa, a wide range of trees and shrubs may be used in contour strips. Some likely species are listed in Appendix I. Trees which grow tall and become part of the upper and middle stories of contour strips should be fast growing, should produce a minimum of shade and root competition and should provide useful products such as fruits, wood or fodder. In all cases, grasses or other herbaceous cover crops should be included. Leafy vegetables that grow readily along hedges or fences may also be included if the field is protected from animals.

A diverse mix of small, dense shrubs and herbaceous plants in the understorey is also important. This may include fodder plants such as

Dichrostachys cinerea along the edge if controlled browsing is planned. *Stylosanthes, Crotalaria* and *Indigofera* species and *Lablab purpureus* all contribute to soil fertility, ground cover and fodder production. Natural vegetation can be encouraged within these plantings or simply preserved and supplemented later if needed. Grasses useful for the understorey include *Pennesetum purpureum, P. typhoides* (elephant grass), ITripsacum laxum (Guatemala grass), *Panicum coloratum, P. maximum* (Guinea grass), *Chloris gayana* (Rhodes grass), *Cynodon dactylon, Cenchrus ciliaris, Dichanthium annulatum* (Starr grass, used in India) and *Eragrostis* species.

Contour strips may be established by intensive planting or simply by maintaining strips of vegetation when clearing new fields, particularly where natural vegetation consists of bush thickets, thickly wooded savannah or *miombo* woodlands. This technique has the advantages of requiring little labour, while also preserving some indigenous woodland and bushland for private use by farmers. The disadvantages include preserving possible refuges for weeds and pests close to croplands and possibly the need to fence the wooded areas for controlled grazing, usually at considerable expense.

In densely populated regions with intensive farming systems, strips are usually planted with a combination of seed, cuttings and seedlings. The exact order of planting and the time required for the vegetation to mature vary by region and with each combination of species and pattern of placement. Generally, grasses, other groundcover, shrubs and trees should not all be established at the same time: The grasses should usually be planted first to form a continuous barrier to trap runoff.

Women harvesting fruit from a mature contour strip that combines trees, shrubs, grass and vegetables.

MANAGEMENT

Management of contour strips may vary from intensive hedge lopping and grass and fodder cutting to occasionally harvesting tree products or allowing animals to graze on the strips. Normally, livestock may be allowed to graze the strips along with stubble after field crops have been harvested. Special protection from animals may not be required if the larger cropland plot is already well fenced. If this is the case, contour strips are an ideal location to grow cut-and-carry fodder.

Regardless of the specific design of contour strips, a continuous and dense understorey of groundcover must always be maintained. This lower layer must be dense enough to trap and hold soil, leaves and stick litter. In order to function properly, contour strips must slow down runoff water and allow it to soak into the soil. Any gaps in the vegetation will concentrate runoff, resulting in damage downslope. Plant litter such as leaves and small branches may be wedged between stems and trunks at ground level to increase the barrier's effectiveness, although in some regions this may attract termites.

Depending on slope and type of soil, trapped water and sediment may harm the vegetation growing in contour strips. Apart from water logging, accumulated sediments may choke or bury young, sensitive plants. The health of the vegetation should be checked regularly, with replacement by hardier plants where necessary.

ANTICIPATED BENEFITS

Men planting tree seedlings along a grass strip.

In addition to tree products, such as food, fuelwood, and fodder for livestock or bees, leaf litter from contour strips can add organic mat-

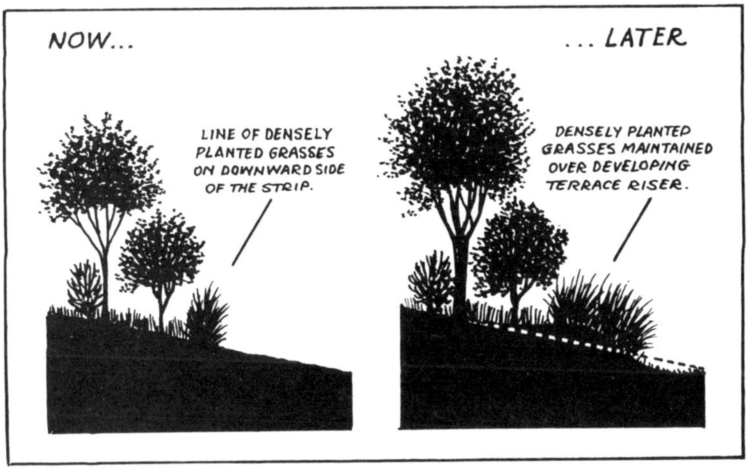

ter to the adjacent cropland as it decomposes, particularly if dug or tilled into the soil. If left on the surface, leaf litter will protect the soil from erosion. Plants may be chosen for contour strips which improve the soil by fixing nitrogen and bringing nutrients up from the subsoil and contour strips may also serve as windbreaks.

In general, farmers are most likely to select fruit or fodder trees for contour strips that do not grow too tall or that can be cut frequently. From a single 100 x 1 metre strip, a farmer in the savannah zone can expect to harvest 200 kg dryweight of fuelwood and 250 kg of fodder a year. Five such strips could produce 25% of the annual fuelwood requirements for a family of five and enough fodder to feed one cow for 8 months of the year.

However, a proportion of cropland is inevitably taken up by contour strips and crop yields may decrease correspondingly in the short term. The diverse products of the trees and shrubs, in addition to their contribution to soil and water conservation, will offset this short-term decline in crop production. The importance of these benefits depends on the severity of soil and water loss and the importance of local requirements for fuelwood, fodder and other tree products. Local community members should be involved in deciding what trade-offs and risks they are prepared to take.

EXAMPLES FROM THE FIELD

In a project in Lushoto, Tanzania, farmers do not practice terracing, but rather combine rows of pineapple, grasses and trees in strips to conserve the soil on sloping fields of maize and beans. In Morocco under very different conditions, *Tamarix* species have been widely planted in vegetation barriers of various shapes for land reclamation and dune stabilization.

One of the best-documented agroforestry systems of this type has been developed in the Nyabisindu region of Rwanda under fairly good soil and climate conditions. In the understorey, grass strips alternate with leguminous groundcover (*Desmodium* species). The upper storey includes *Grevillea robusta*, *Albizia* species, *Leucaena leucocephala* and *Entada abyssinica*.

SELECTED REFERENCES

Anger, G., ed. (1985). *L'arbre et la haie dans l'exploitation paysanne.* Fiche Technique No. 3. Nyabisindu, Rwanda: Project Agropastoral de Nyabisindu, 126 pp.

Douglas, M.G. (1988). *Integrating conservation into the farming system: land-use planning for smallholder farmers—concepts and procedures.* London: Commonwealth Secretariat, 130 pp.

ICRAF (1987). Agroforestry for the Control of Soil Erosion. *ICRAF Newsletter.* 19: 1-2.

Young, A. (1986). *The potential of agroforestry for soil conservation. Part I: erosion control.* Working Paper No. 42. Nairobi: ICRAF, 89pp.

4.3 Alley Cropping

DESCRIPTION

Alley cropping, also known as hedgerow intercropping, involves managing rows of woody plants with annual crops planted in alleys in between. The woody plants are cut regularly and leaves and twigs are used as mulch on the cropped alleys in order to reduce evaporation from the soil surface, suppress weeds and/or add nutrients and organic matter to the topsoil. Where nitrogen is required for crop production, nitrogen-fixing plants might be the main components of the hedgerows.

One form of alley cropping.

The primary purpose of alley cropping is to maintain or increase crop yields by improvement of the soil and microclimate and weed control. Farmers may also obtain tree products from the hedgerows—including fuelwood, building poles, food, medicine and fodder—and on sloping land, the hedgerows and prunings may help to control erosion.

Alley cropping is designed to be a sustainable alternative to shifting cultivation or expansion into unproductive farmland. It retains the basic principles of traditional fallowing, but keeps all the land productive at the same time. Alley cropping usually works best in places where people feel a need to intensify crop production but face soil fertility problems. This situation is often characteristic of crowded, densely populated areas, but may also occur wherever some farmers wish, or are forced, to increase production on a plot of limited size.

Alley cropping is distinct from other agroforestry practices, such as contour vegetation strips (see section 4.2), which might look similar, because with alley cropping the focus is on soil improvement by mulching between the hedgerows. Contour vegetation strips, by contrast, are designed to reduce the length of slopes and the speed and amount of water flowing across the soil surface. Alley cropping may serve both purposes if specifically designed to do so, but the emphasis is usually primarily on cycling nutrients and producing mulch and only secondarily on providing a live barrier for soil and water conservation.

Alley cropping also differs from planting woody and herbaceous plants in ditches, ridges and terraces (see section 5.1) because it does not usually require physical structures; neither is it intended to stabilize structures nor to produce useful goods on the space taken up by structures. Alley cropping can serve these purposes in addition to its primary function if specially designed to do so.

Alley cropping is new to most of dryland Africa: Most trials have been conducted in the humid and subhumid areas of the continent. Planting closely spaced lines of trees in cropland is not traditional in rural areas, with very few exceptions. However, important conservation and production benefits have been reported at sites where alley cropping has been introduced on a trial basis.

DESIGN

Woody plants are introduced as hedgerows in farm fields to maximize the positive and minimize the negative effects of trees on crop management and yields. Without a doubt, trees compete with farm crops for soil nutrients, soil moisture and light. However, the right kind of trees at the right spacing, with proper management, may actually produce a net increase in yields from cropland. Trees may also provide new products, such as fuelwood, fodder or food, in addition to the annual crops. In some cases, the introduction of alley cropping is justified by immediate improvements in crop yields, while in other cases—perhaps the majority—alley cropping is justified by long-term improvements in soil fertility and sustainability.

The position and spacing of hedgerow and crop plants in an alley cropping system depend on plant species, climate, slope, soil conditions and the space required for the movement of people and tillage equipment. Ideally, hedgerows should be positioned in an east-west direction so that plants on both sides receive full sunlight during the day. The spacing used in field trials usually ranges from 4 to 8 metres between rows and from 25 cm to 2 meters between trees within rows.

The closer spacing is generally used in humid areas and the wider spacing in subhumid or semiarid regions.

Where farmers do not favour such a close association of crops with hedgerows, the same number of hedgerow plants might be arranged in wider (double) hedgerows spaced further apart. This may lead to more competition between hedgerow plants, but will reduce direct competition between the hedgerow and the crops. This spacing can also be used to accommodate higher hedgerows with larger trees interspersed.

Position and spacing of hedgerows may also be affected by slope and the placement and design of soil and water conservation structures, where these are combined with alley cropping. On sloping land,

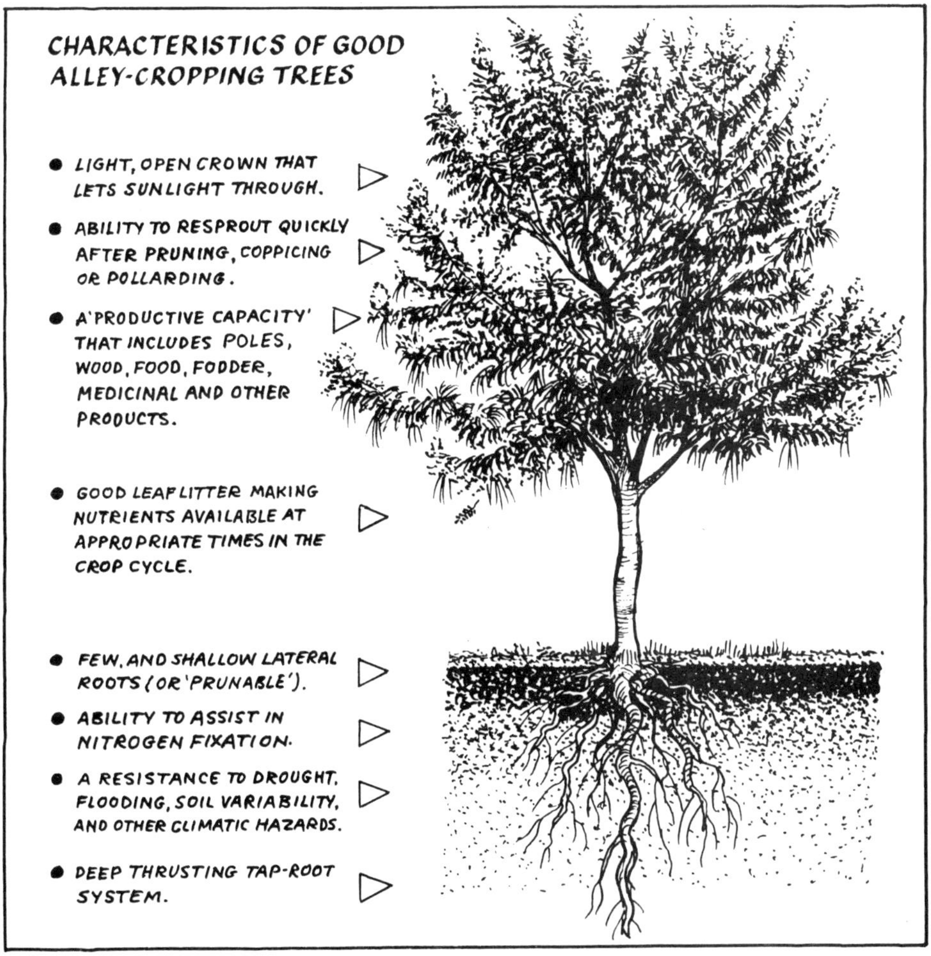

hedgerows should always be placed on the contour. If this means that they do not have the desirable east-west orientation, then they may need regular trimming to prevent excessive shading of adjacent crops.

SPECIES

Alley cropping usually includes leguminous trees to improve soil fertility through nitrogen fixation. Apart from this important trait, some woody species are obviously less compatible with crops than others. Large dense crowns and surface roots extending far beyond the crown line interfere with most crops. Trees that use a lot of water (high evapotranspiration) may also reduce the water available for the crops. Some trees and shrubs are poor candidates for any agroforestry practice in cropland because they give off toxic chemicals into the surrounding soil, a defense against weeds and other competitors in the forest environment. Even woody species that are otherwise compatible with crops are often not suited to alley cropping because of their size or poor response to frequent pruning.

An ideal alley-cropping tree or shrub should have several characteristics. It should have a sparse, small crown to permit sunlight penetration or should resprout rapidly after pruning, coppicing, pollarding or lopping. It should form a deep tap-root system with few lateral root branches near the surface, so as not to compete with crop roots. Alternatively, trees or shrubs may be used with shallow lateral roots that are easily 'pruned' by ploughing along the hedgerow, without serious damage to the plants. The leaf litter, or some portion of it, should decompose at a rate that makes nutrients available when they are needed in the cropping cycle. Ideally, trees and shrubs used for alley cropping should fix nitrogen and should also produce wood, food, fodder, medicine or other products used by farmers or other members of the local community. Finally, the species selected should grow well under the specific limitations of the site, such as saline or acid soils, drought, flooding, heavy winds, insect pests or other hazards.

Many alley-cropping projects have used *Leucaena leucocephala* because of widespread reports of high yields and nutrient cycling capability. However, the performance of this species varies widely according to climate and soil type. *Leucaena leucocephala* has often failed, and even harmed neighbouring crops, because it attracts animal and insect pests, including domestic stock, small antelope and termites. Under continuous browsing, the hedgerow plants may survive but never grow taller than 10 to 20 cm. When termites attack, they often succumb leaving no trace except perhaps some damaged crops nearby. A sap-sucking insect pest, *Heteropsylla cubana*, that causes widespread damage in Asia has now been reported in Africa and may threaten continued, large-scale use of *Leucaena*. This shows the impor-

tance of maintaining species diversity in agroforestry systems. One alternative to *Leucaena* for alley cropping is *Gliricidia sepium*.

The long list of performance criteria, combined with the example of potential problems associated with the overuse of one species, shows how difficult it is to choose one 'ideal' alley cropping species or mix of species for all places and situations. The criteria listed here can only serve as a guide for species selection at particular sites. In some cases, it may be necessary to combine two or more species in the hedgerow to provide the products and benefits desired.

For example, in the dry savannah region of Kenya some trials have combined one species with fast-decomposing leaves, to release nitrogen quickly into the soil, and a second species with longer-lasting leaves, to maintain soil cover throughout the entire cropping season and to build up organic matter in the top soil. While neither of the species used—*L. leucocephala* and *Cassia siamea* —proved to be fully suitable in this area, the idea of combining two species with different leaf decomposition rates might usefully be applied in many sites.

Cassia siamea, Gliricidia sepium, Calliandra calothyrsus and *Sesbania sesban* are commonly used tree species for alley cropping. For an initial list of tree species suitable for a particular area, see Appendix I and check your own list of locally available trees against the requirements of species which can be used for alley cropping.

MANAGEMENT

Management practices for alley cropping vary widely in specific locations. Hedgerows may be established by direct seeding or by planting

seedlings or cuttings. Seedlings may be placed in a deep ploughed furrow or in deep individual planting holes. During establishment, the hedgerow may require protection from browsing animals, trampling or pests. If the hedges are sown or planted along with crops, they will be protected as the farmer protects the field as a whole. The young trees will benefit from weeding, fertilizing and whatever management the farmer follows for the crops. These activities will encourage growth of the hedgerows so that later they will require little additional attention.

Once established, trees and shrubs used in alley cropping are usually left to grow for 6 to 18 months before the first cutting. The timing depends on the vigour, root development and height of the hedgerow plants and on the type of hedgerow and cutting schedule required. Most decisions about managing hedgerows are a compromise between keeping the woody plants in good condition for long-term production and avoiding short-term damage, loss or problems with existing crops.

The first cutting should be late enough to allow the woody plants sufficient root development and resilience, yet soon enough to avoid shading adjacent crops. In subhumid areas or in semi-arid areas with high production potential, hedgerows can be cut within 6 to 12 months of planting, whereas in drier areas cutting should usually be delayed for 12 to 18 months after planning or even longer. Some fruit or timber trees, dispersed in the hedgerow, may be allowed to grow to full size with little or no pruning.

The type and frequency of cutting must also be adjusted to specific site conditions. The hedgerow plants may be coppiced, pruned, pollarded or lopped into a variety of shapes and sizes. The choice depends on several factors, including the crop and hedgerow species, the relative importance and type of products, by-products and services expected from the hedgerows and the amount and timing of labour available for hedgerow management and harvesting.

A standard alley-cropping practice consists of coppicing hedgerow plants at a height of 30 to 60 cm, followed by lopping to the same height at intervals ranging from once a month (during the cropping season) to once a year (around the beginning of the cropping season). Tree-leaf mulch is usually applied just after land preparation or crop sowing. Most formal alley cropping experiments have been conducted at research stations in the humid lowlands, so most published recommendations reflect an emphasis on hedgerow management to produce as much mulch as possible, properly timed for the best effect on crop yields. These recommendations may not apply as well in dry regions or under actual conditions in farmers' fields, so they should be followed with a degree of caution.

Smallholder farmers in dry areas have priorities and limitations that may affect their approach to hedgerow management. For example, one farmer might decide to prune some trees on the sides to encourage rapid growth of tall trunks with small, high canopies. Trimmings can be applied as mulch, poles can be harvested after reaching a useful size, usually in 4 to 10 years, and, after coppicing, the tree can be lopped for leaf mulch or pruned for more poles. Another farmer might choose to allow tethered goats to browse on regrowth during the dry season, then skip a season and coppice the hedgerow regrowth for mulch in the following cropping season. As another consideration, farmers may vary the timing of hedgerow management tasks to fit in with ploughing or weeding schedules.

In areas where domestic or wild animals damage the trees in hedgerows, farmers have suggested planting trees in small blocks close to the home. While the leaf mulch then has to be carried to the fields, only a small block of trees has to be fenced. Farmers have tested and will continue to develop many other variations on the standard pattern of alley cropping.

ANTICIPATED BENEFITS

Alley cropping in dryland Africa is perhaps the least proven of all the agroforestry practices presented in this handbook. However alley cropping has the potential to make many traditional and emerging cropping systems more productive on a sustainable basis. Trials and experience in the field will show which types of alley cropping are most productive in specific situations, for instance to help farmers maintain the produc-

tivity of their fields under continuous cultivation, rather than relying on long fallows, slash and burn techniques, fertilizers or pesticides.

Numerous trials and experiments are in progress focussing on alley cropping in different regions of Africa. Preliminary results are still incomplete and often site specific. These have stimulated a certain amount of speculation.

As a result of improved soil fertility, crop yields under alley cropping have dramatically increased in field trials conducted by the International Institute of Tropical Agriculture (IITA) in the subhumid zone of Nigeria. *Leucaena leucocephala* and *Gliricidia sepium* were used by farmers to provide living stakes for yams in a maize-yam alley cropping system. Significant increases in maize yields were observed when *Leucaena* prunings were added to the alleys as green manure. Ten tons of prunings, dug into the soil at the time of maize seeding, increased grain yields from 1.3 to 3.2 tons per hectare. This is reportedly equivalent to applying 100 kg of nitrogen fertilizer per hectare.

Several alley-cropping projects in Kenya have reported mixed results. At this point, it would be premature to launch a large-scale, extensive promotion of alley cropping in other parts of Africa on the basis of the limited information available. Extension efforts should be restricted to specific areas where experiments and farmers' experience have shown major benefits from a given alley-cropping system.

As part of a CARE project in Kenya's Siaya District, farmers have introduced a few lines of trees in their fields; if results are encouraging, they can plant more. This incremental approach limits the level of risk, spreads labour requirements over a long period and lets farmers see for themselves whether alley cropping is worthwhile under their production conditions.

Most often, the best approach is to promote alley cropping as a technique to be tested and modified under local conditions. Such efforts are probably best restricted to small-scale pilot projects until farmers and researchers feel more certain of the expected costs and benefits of alley cropping in a particular region. Where large numbers of farmers are willing to experiment, researchers may wish to extend small alley-cropping trials to a wider group, though still on an experimental basis.

COMBINATION WITH OTHER TECHNIQUES

In many cases, a combination of hedgerows with dispersed trees in cropland can provide additional or better products and a greater impact on surrounding crops than a simple alley-cropping system. Farmers may wish to combine alley cropping with carefully spaced individual fodder trees, such as *Acacia albida*, fruit trees, such as *Persea americana* (avocado) and *Carica papaya*, or trees intended for pole production. Hedgerows may provide a site for individual trees that serve a purpose very different from that of the hedgerow itself.

Alley cropping can also complement contour vegetation strips (see section 4.2) and structural measures for soil and water conservation (section 5.1). In such cases, the position of the hedgerows follows, first, the placement of the structures or contour strips and, secondly, the guidelines for alley cropping. Mulch production complements the erosion-control function, while the hedgerow plants strengthen conservation structures and improve soil fertility in the surrounding fields.

Farmers have combined alley cropping with many other practices. For example, farmers participating in agroforestry research projects in Kenya have expressed an interest in combining tree litter from hedgerows, blocks or fencelines with composting or related techniques. In one case, farmers reported adding leaves and twigs of *Euphorbia tirucalli, Terminalia brownii* and *Combretum* species to cattle pens for composting. They were interested, not in the structure of alley cropping, but in the idea of nutrient cycling by adding leaf litter to the soil.

Many farmers with livestock also see alley cropping as an important complementary source of fodder for stall feeding or controlled browsing during the dry season. Fodder from hedgerows may supplement fodder available from tree lots, improved fallow (section 4.5) or woody browse plants in pastures (Chapter 7). In this case the structure of alley cropping is adopted but the primary purpose is changed.

EXAMPLES FROM THE FIELD

Although alley cropping is fairly new in Africa, related practices have been promoted as part of soil conservation efforts. The best-known series of alley-cropping trials in Africa is being conducted by IITA and the International Livestock Centre for Africa (ILCA) at several locations in Nigeria, where a number of alley-cropping systems are being assessed. Researchers have gathered detailed information on mulch, wood, fodder and crop production, and on site improvement and labour requirements. Many research stations in Africa are now repeating some of IITA's alley-cropping experiments to collect information on the practice under a wider range of site conditions. The Soil Productivity Research Project at Misamfu Research Station in Zambia (*miombo* woodland/acid soils) and the Dryland Agroforestry Research Project in Machakos, Kenya, are testing alley-cropping techniques suitable for dry areas.

Important alley-cropping experiments, trials and extension efforts have been conducted in Rwanda under the Direction General des Forêts and GTZ. In addition, a number of special projects have included alley cropping trials in recent years. The Kenya Renewable Energy Development Project (KREDP) conducted alley-cropping trials in four different agroclimatic zones, testing different spacing, species and management. At the Kenya coast, with annual rainfall varying from 600 to over 2000 mm, *Acacia albida, Gliricidia sepium, Casuarina equisetifolia, Adenanthera pavonina* and other species were used in alley-cropping trials on acidic, infertile, sandy soils. Yields of maize and green gram increased as much as 60% after 3 years due to improvement of soil fertility. Wood yield from *Casuarina* was as much as 86 cubic metres per hectare and weed control improved by up to 80%; improved weed control may be the most significant benefit for some farmers.

ICRAF and the Kenya Forestry Research Institute (KEFRI) have initiated several small trials to test alley-cropping research methods for research stations and on-farm plots. Observations and limited measurements at semi-arid sites in Kenya's Machakos District indicate a wide variation in the effects of hedgerows on crops, depending on soil, slope, climate and management. Farmers tend to prefer other spacing arrangements, such as trees dispersed in cropland at 2 x 2-metre intervals or widely spaced hedgerows along terrace risers with 8-meter intervals between rows. The Dryland Agroforestry Project in Machakos has experimented with alley cropping using *Terminalia brownii, Cassia siamea* and *Leucaena leucocephala* with maize and with beans. These trials are still in progress.

The CARE-Kenya Agroforestry Project in subhumid areas of Kenya's Siaya District has documented traditional practices that incorporate various aspects of alley cropping. Several farmers in this project area are now testing alley cropping on a trial basis. They have been particularly interested in the reduction of *Striga* weed on some plots. Measurement of yields and site improvement are in progress. Similar trials are being conducted at many other places in dryland Africa.

SELECTED REFERENCES

Arap Sang, F.K., Hoekstra, D.A. and Okumu , R. (1985). *Preliminary results of on-station green manure trials with leaves from* Leucaena leucocephala, Cassia siamea *and* Terminalia brownii. Research Report No. 1. Kikuyu, Kenya: KEFRI 21pp.

Amare Getahun and Bashir Jama (1986). *Alley farming in the coastal area of Kenya.* Nairobi: Ministry of Energy, Kenya Renewable Energy Development Project, 26pp.

Atta-Krah, A.N. and Francis, P.A. (1987). The role of on-farm trials in the evaluation of composite technologies: the case of alley cropping in southern Nigeria. *Agricultural Systems.* 23: 133-52.

ILCA (1986). *Panesa Newsletter* (Pasture Network for Eastern and Southern Africa). Addis Ababa: ILCA.

Tanzania, Ministry of Lands, Natural Resources and Tourism (1984). *Trees for village forestry.* Dar es Salaam: Ministry of Lands, Natural Resources and Tourism, Forest Division, 125pp.

Kang, B.T. and Duguma, B. (1984). *Nitrogen management in alley cropping system.* Ibadan, Nigeria: IITA, 18pp.

Sumberg, J. and Okali, C. (1988). Farmers, on-farm research and the development of new technology. *Experimental Agriculture.* 24: 333-42.

Wilson, G. and Kang, B.T. (1981). Developing stable and productive biological cropping systems for the humid tropics. In B. Stonehouse, ed. *Biological husbandry: a scientific approach to organic farming.* London: Butterworth, 193-203.

Yamoah, C.F., Ay, P. and Agboola, A.A. (1986). The use of *Gliricidia sepium* for alley cropping in the southern guinea savanna zone of Nigeria. *International Tree Crops Journal.* 3: 267-79.

4.4 Trees in Home Gardens

DESCRIPTION

Several kinds of home garden in the drier regions of Africa and other continents include trees and shrubs. These consist of diverse mixtures of vegetables, fruits, medicinal plants and often fodder grasses, shrubs or trees in small, intensively cultivated plots in and around home compounds. Home gardens have a special role in screening new agroforestry and conservation practices. They are among the most suitable sites for experimentation with new plant species, combinations, spacing or management. Since the agroforestry techniques described in this manual originated from a wide range of conditions and environments, it is wise for farmers to test them, make modifications, observe the results and experiment in a place that is convenient, well protected and with some water available—such as a home garden.

In the dry farming areas of Kenya, home gardens may be small vegetable plots with *Passiflora edulis* (passion fruit) vine on the fence and a few *Psidium guajava* (guava) and *Citrus* trees in the overstorey. The traditional home gardens on the slopes of Mount Kilamanjaro in Tanzania, as well as in Cameroon and the Comoros, have three or four storeys, with timber, fruit, fodder, fibre and fuelwood trees over spices, herbs and vegetable crops.

The decision to intensify production in the limited area of a home garden, rather than on cropland, depends on many local factors. For

A Chagga home garden on the slopes of Mt. Kilimanjaro.

Tending a Hausa home garden.

example, in northeast Zambia on the acid soils of the *miombo* woodlands, farmers experiment in home gardens and river floodplain gardens to improve their diet and to earn cash. Despite extensive surrounding woodlands, these farmers have concentrated their efforts on land with access to water, roads and markets. Rather than moving to the outlying woodlands where they could expand their holdings, they are intensifying production on the limited land which they already possess and, in many cases, are trying to define and secure their land rights. While conventional wisdom suggests that home gardens are best suited to densely populated areas where land is in short supply, this case demonstrates that farmers may be motivated to adapt and develop multistorey production systems in home gardens even in situations where they have access to large tracts of land.

Home gardens increase productivity because they are labour intensive, yet, because they are near the home, the labour required can be combined with home and child-care responsibilities. In semi-arid regions of Kenya, women are increasingly left to farm on their own due to the widespread migration of men to cities. Since men's labour for clearing and ploughing is often lacking, the women find it more productive to farm intensively on small, protected plots.

Women have been key participants in agroforestry projects throughout Africa. Although home gardens may be managed by either sex, they are most often managed by women. Home gardens provide a legitimate place for women to cultivate agricultural crops since they are usually located close to the home compound and are seen as an extension of the home. This is particularly important in areas of North Africa and the Sahel where women do not traditionally till the land. Here and in other areas, home gardens are accessible to women whose mobility may be limited by custom, or by responsibilities for child care, food processing and preparation.

Intensive home gardens are also particularly useful for women who are wage labourers and heads of household. These women can perform agricultural tasks throughout the season and they can harvest important foods for home use and for sale without animal or mechanical draught power.

Home gardens are often the best place to initiate agroforestry projects with women. They are also an ideal site for introducing soil and water conservation measures. The limited plot size encourages multistoreyed systems, while the degree of control and permanence associated with the home site encourages investment in tree crops and other site improvements, such as earthworks, irrigation, manuring and fencing. The small plot size implies a high ratio of boundary area to enclosed space, so that multipurpose living fences may provide a large proportion of production. Home gardens can also accommodate small livestock, such as chickens or rabbits, and may provide residues or fodder for pigs, goats or dairy cows. The home garden, like the home compound as a whole, may also serve as a testing and observation plot for new species of plants or for cultivating plants that were previously gathered.

The home garden also has a valuable role in educating children about agroforestry. Children are often responsible for carrying water and doing other useful maintenance work and they may be given a few plants or a small area to tend themselves. Like establishing tree nurseries in schools, this process involves the next generation in improving the management of their land.

Many home gardens do not yet include trees or other woody plants as major features. However, intensive techniques of intercropping, fencing, occasional watering (using waste water or water harvesting from the compound), soil enrichment and pest management, which may already be used in home gardens, provide the basis for developing multistoreyed agroforestry systems. The combination of intensive gardening with water conservation and harvesting may also increase the value of the plot so that livestock are excluded, perhaps by 'social fencing' (i.e. by custom or agreement). Trees can be planted on the site once its value, and thus its protection from animals, are established.

DESIGN

Field workers may wish to encourage the use of home gardens in a number of different situations. Where tree products that used to be gathered off the farm are now in short supply, they can be cultivated in a home garden. The introduction of woody species may also provide new products for home consumption and sale. In situations where the

Watering a home tree nursery in Machakos District, Kenya.

limits of intensified production have already been reached, farmers may be prepared to extend multistorey garden practices onto their main cropland, especially if technical and marketing assistance is available. As with other agroforestry practices, the needs and priorities of the local people provide the basis for the design of home gardens.

To create a multistorey home garden, several approaches are possible. In some cases, people may add new tree species to existing gardens; in other cases they may add vegetables, fruits or rootcrops beneath an open canopy of existing trees. Yet other farmers may start afresh, without prior gardening experience, on a clean-tilled plot. Whichever approach is taken, farmers eventually need to make management decisions which will affect their other production systems, such as whether to use organic and/or inorganic fertilizer on garden plots or whether and how to collect and manage water for gardens. The establishment and maintenance of a home garden may require significant amounts of time and labour.

At the outset, the farmer must choose the size of the plot and decide whether and how to fence it. Animals must be kept out of the garden, whether by temporary or permanent fencing. If the plot size may change, then part of the fence should be temporary, subject to future decisions to expand.

If many trees are to be planted, farmers must also decide how extensively and how deeply to dig or fill the land, whether to use raised or sunken beds, a uniform clean till or surface mounds. These decisions should be based on the site's drainage and soil fertility, the requirements of the crops and the value of potential harvest versus the labour required for site improvements to increase yields.

In the *miombo* woodlands of northern Zambia, women heading households and subhouseholds see woodland gathering and home gardens as their best strategies for supplementing family food supplies and cash income. They may use mounded soil in gardens ranging from 5 x 5-metre plots scattered around the homestead to continuous quarter-hectare plots surrounding the house. Crops usually include some combination of tobacco (if men are at home), cassava, pumpkins, gourds, sweet potatoes, red beans, guava, citrus, castor beans and Ceara rubber (*Manihot glaziovii*).

SPECIES

To plan a home garden, farmers need access to seeds and seedlings. They also need accurate information on a wide variety of species, including new and familiar grasses, fruits, vegetables and multipurpose trees. Community members may choose to collect and cultivate wild species which were previously gathered, for example in Kenya's Machakos and Bungoma Districts and in Zvishavane District in Zimbabwe.

Home gardeners need information on the shape and size of woody species under various types of management in order to determine the best spacing of trees and shrubs. Agroforestry and social forestry projects often need to collect and disseminate information on indigenous trees which may be well known but have not previously been cultivated or managed intensively.

Technical-assistance personnel need to provide information on soil conditions, nutrients and the amount of water and light required by woody plants and understorey crops. Farmers need to know about each species's vulnerability to insect pests or browsing animals, as well as its ability to withstand drought, waterlogging or shallow, rocky, crusting or cracking soils. Farmers also need information on potential competition between various species and some examples of productive species combinations.

Finally, farmers want information on potential yields, current and potential market values of products and the amount of labour and other inputs—such as manure, mulch, chemical fertilizer or insecticide—required for a plot of a given size. The most accurate information on these topics is obtained from experience at the site. Farmers may start with a rainfed (non-irrigated), low-input garden and proceed in steps towards more intensive cultivation.

Pictures or local examples may spark informative discussions about what products people expect from home gardens. In many cases, farmers prefer to maximize economic value, rather than biologically ideal combinations and spacings. Farmers may also choose to include fibre, medicinal or spice plants of special value for domestic use, even

Sesbania sesban *supporting passion fruit vines.*

if they are not ideal for combination with fruit, timber or fodder trees. It is important that people have the information they need to make decisions according to their own conditions and needs, rather than copying some 'ideal' multistorey garden.

Plants that give off unpleasant odors or irritating pollen should be avoided, as well as species or varieties that will rapidly outgrow the garden. For example, one of the most common requests from experienced home gardeners is for dwarf fruit trees.

While the list of appropriate species for home gardens is virtually unlimited, some of the most frequently used species are:

Woody	Herbaceous
Anacardium occidentale (cashew)	*Abelmoschus esculentus* (okra)
Artocarpus heterophyllus (breadfruit)	*Allium* species (onion)
Carica papaya (papaya)	*Arachis hypogaea* (groundnut)
Citrus species	*Brassica oleracea* (cabbage)
Manihot esculenta (cassava)	*Cucurbita* species (pumpkin)
Passiflora edulis (passion fruit)	*Ipomoea batatas* (sweet potato)
Persea americana (avocado)	*Lycopersicon esculentum* (tomato)
Phoenix dactylifera (date)	*Maranta arundinacea* (arrowroot)
Psidium guajava (guava)	*Musa* species (banana, plantain)
	Phaseolus species (beans)

Most well-developed home gardens also include medicinal plants, condiments and indigenous fruits and vegetables. A particularly creative combination devised in Western Kenya uses *Sesbania sesban* as the support for *Passiflora edulis* (passion fruit) vines.

ANTICIPATED BENEFITS

Potential benefits from home gardens are as varied as the gardens themselves. These may include cash earned from direct sale of produce, cash savings realized by substituting garden produce for purchased vegetables and improved nutrition from new types of food or from increased, regular quantities of vitamins, minerals and plant protein in the household diet.

The Chagga people cultivate home gardens on the slopes of Mount Kilamanjaro in Tanzania, with an annual rainfall averaging 1000 to 1700 mm. On a typical plot of 0.68 hectares, farmers produce about 125 kg of beans (148 kg/ha), 280 kg of unhusked coffee (412 kg/ha) and 275 bunches of bananas (404/ha). Additional fruits, vegetables and herbs, grown mainly for home consumption, have not been quantified. Farmers keep from three to five traditional beehives, each producing at least 5 kg of honey annually.

The Chagga farmers are also almost self-sufficient in fodder, produced primarily from the trees, shrubs, banana plants and grasses grown in home gardens and used to stall-feed their livestock.. Typical livestock holdings include three cows, two goats and six chickens. It is estimated that each home garden also supplies from 1 to 2 cubic metres of fuelwood per year (1.5-3.0 cubic metres/ha), which provides one-quarter to one-third of a household's annual fuelwood requirements.

Although local sources indicate that coffee and/or maize or bean crops fail once every 3 to 4 years, there has never been a failure involving all of the products of the Chagga home-garden system. Thus the risk of crop failure is reduced by cultivating a diversity of products.

Home gardens are convenient, providing easy access to leaf protein, condiments, medicinal plants and shrubs so that women and children do not have to make long gathering trips. While such trips can be pleasant, the gatherers may prefer to use the time for other activities. In addition, they can control the household site and its use more strictly than they can control shared land or land on the periphery of their own larger holdings. Home gardeners can extend the harvesting season by soil and water management, and by protecting the plants from harvesting by others or grazing by animals.

In terms of soil and water conservation, home gardens provide the setting for training and experimentation with new, more intensive management, with new tree species and with new uses of familiar species. Techniques developed in home gardens may eventually be applied in cropland, rangeland, river banks, reclaimed gullies or flood plains.

The value, rather than the quantity, of home-garden products is an important consideration. For example, a woman smallholder with limited land and labour may double the value of her products by changing to labour-intensive cultivation of a few valuable crops on a small garden plot. Likewise, households with no access to shared land for collecting wild vegetables, fruit, timber, fibres or other useful plant products may benefit substantially from cultivating previously gathered plant species in a small garden. These products may have special importance far beyond their market value if the household lacks an alternative source of supply.

COMBINATION WITH OTHER TECHNIQUES

Home gardens usually need to be completely enclosed by some sort of fencing to keep out livestock and wild animals. For this reason, a small, tightly woven living fence (see section 6.1) is a useful complement to most home gardens, especially in areas where animals are a major hazard for tree seedlings and vegetables crops. Such fences can also serve as windbreaks (section 6.3). Multistorey gardens may be established in areas reclaimed and stabilized using other agroforestry techniques, for instance in gully sites, flood plains, along stream banks or near seasonally flooded depressions (sections 5.3 and 6.4).

Although home gardens are primarily used among settled farmers, women in pastoralist groups may keep a few livestock, such as young or sick animals or animals in milk, in special small enclosures close to the home. This is the case, for example, among the Maasai in southern Kenya. The location minimizes time away from home and allows for a special investment of time, attention and protection. This might be a reasonable situation in which to introduce more intensive agroforestry practices for fodder production.

Members of a community group tend tree seedlings in a horticultural nursery.

Home gardens may also provide a site for community or social forestry projects. Women may choose to plant and maintain trees for fuelwood or other products in a home garden or a block near their homes. Women's groups in some areas have combined vegetable gardens with tree nurseries at a community site. This idea could be expanded to combine multipurpose trees with horticultural and vegetable crops, rather than a single-species stand of timber or fuelwood trees, in a large community plot close to women's homes.

Timber or fuelwood species compatible with multistoreyed, multipurpose home gardens could help to alleviate conflicts associated with men's timber versus women's fuelwood production, as reported in Kakamega District, Kenya; or between men's versus women's land-use priorities, as reported in Kenya's Kisii District; or between commercial tree production versus subsistence agriculture, as reported from social forestry projects in India. Establishing valuable horticultural crops within a community tree plot can help to identify the site as one of value, thus strengthening 'social fences' to prevent grazing and browsing.

EXAMPLES FROM THE FIELD

Successful home-garden systems are well known throughout the humid lowlands of Africa. Less well known are the more experimental systems being developed in the savannah and dry woodland zones.

For example, in the *miombo* woodlands of northeastern Zambia, women's home gardens are becoming increasingly important for food production and cash income and some women are experimenting with the introduction of fruit and other trees. Farmers are testing various agronomic practices, such as mounding, raised beds and clean tilled plots, with a preference for mounding in the larger gardens as a way of incorporating grass with some tree and shrub parts into the soil. The mounding of loose topsoil over plant biomass has been adapted from the grass-mounding techniques of a neighbouring community for the cultivation of beans, cassava, fruits, vegetables and other crops in women's home gardens.

Women who are heads of households rely largely on cassava (*Manihot esculenta*) production in home gardens to supplement the food they buy with wages. In households where there is no male labour for land clearing, home-garden production, brewing and cassava processing are important ways to earn cash. These activities reflect a desire to intensify land use on small plots—limited by the lack of male labour, not land—and to diversify economic enterprises.

The Chagga in Tanzania grow food crops—including over 15 types of banana, beans, cowpea, maize, potato, taro and tomato—and cash crops—including coffee and cardamom (*Elettaria cardamomum*)—together with over 40 species of woody plants. The woody species have many uses, from living fences to veterinary medicines to bee forage. The farmers thin the tree canopy to encourage food crops which need more light and to allow seedlings of valuable timber trees to grow into the upper storey. Farmers grow some trees, such as teak, on a 60- to 80-year rotation, and replace cut trees for their descendents. This production system is based on a strong tradition of family land tenure.

Thus, farmers in this area have transformed the indigenous forest into a diverse and productive agroforestry system which has been stable for at least a century. Fuel, fodder and fruit trees have been retained and less useful species replaced with new trees and crops. Although the subhumid climate, with an average annual rainfall of 1000 to 1700 mm, and fertile soils are not typical of dryland Africa, these multistorey gardens illustrate a traditional African agroforestry system which could be adapted for use in other areas.

SELECTED REFERENCES

Bala Subramanian, V. and Agli, A. (1986). The role of agroforestry in the farming systems in Rwanda with special reference to the Bugeresa-Gisaka-Migongo (BGM) region. *Agroforestry Systems.* 4: 271-89.

Fernandes, E.C.M., Oktingati, A. and Maghembe, J. (1984). The Chagga home gardens: a multistoried agroforestry cropping system on Mt. Kilimanjaro (Northern Tanzania). *Agroforestry Systems.* 2: 73-86.

Huxley, P.A. and Rocheleau, D.E. (1985). *Farm systems and agroforestry research in northern Zambia.* Nairobi: ICRAF, 76 pp.

Okafar, J.C. (1980). Edible indigenous woody plants in the rural economy of the Nigerian forest zone. *Forest Ecology Management.* 3:45-55.

Stollen, K.A. (1983). *Peasants and agricultural change in northern Zambia.* Occasional Paper Number 4. As, Norway: Agricultural University of Norway, Department of Agricultural Economics.

Wachiira, K.K. (1987). *Women's use of off-farm and boundary lands: agroforestry potentials.* Nairobi: ICRAF, 137pp.

4.5 Improved Fallows

DESCRIPTION

The practice of leaving cropped fields to lie fallow in order to allow the soil to rest and recover some of its fertility is well known and widely used throughout Africa. When agroforestry is added, woody species replace crops on the fallow fields in sequence over time. The main feature which distinguishes fallow-based agroforestry systems from other agroforestry practices is that trees and shrubs are not grown with crops on the same plot at the same time.

In many parts of Africa, fallow periods are becoming shorter, due to an increasingly acute land shortage. The shorter rotations may fail to restore the soil sufficiently to sustain later crop production. Severe soil losses also occur on many fields in savannah areas during the cropping cycle, which further delays their recovery time when fallowed.

The relatively sparse cover at the beginning of fallow periods, often coupled with heavy grazing, increases the risks of soil erosion. As the fallow period is reduced, it becomes more important to introduce trees and shrubs that could help speed up soil recovery. The best species seem to be those which enhance soil fertility, especially by fixing nitrogen, and which establish ground cover quickly.

Different strategies may be adopted for the introduction of woody plants during the fallow period, and benefits may be expressed in terms

Fallow field planted with Sesbania sesban *as a rotational woodlot.*

Fallow ploughed to leave strips on the contour.

of short-term economic gains or longer-term improvement of soil fertility. Commercial tree crops may be added to the natural regrowth or leguminous fodder and cover crops may be sown. Depleted croplands are sometimes planted with soil-enriching trees, such as *Sesbania sesban*, and managed as woodlots on longer rotations.

Fallow-based systems of soil management are perhaps best developed in the densely settled rainforests and humid savannahs of West Africa. However, fallowing is also practiced in the subhumid *miombo* woodlands of Southern Africa and the dry savannahs and highlands of East Africa. In many cases, the fallow period presents an opportunity to produce useful goods with a minimum of labour. Agroforestry practices are often introduced into fallow management in situations where there is not enough unused land for gathering woodland products, yet still enough land for farmers to maintain a fallow cycle in annual crop production.

The fallow period may range from 1 to 20 years, though the classic bush-fallow systems of West and Central Africa were usually based on an 8- to 10-year period. The optimum duration depends on local criteria such as the immediate need for food or cash crops, the importance of soil fertility improvement and the value placed on crops with long maturation periods, such as timber or fruit trees. Over time, the fallow period will:

- protect the soil from erosion
- eliminate weeds, pests and diseases specific to the cropping system
- increase the organic matter content of the soil, cycle and trap nutrients from the subsoil and improve soil structure, including aeration, water-holding capacity and tilth.

DESIGN

Different approaches can be used to introduce agroforestry practices into a fallow rotation system. Their functions are similar—a combination of site enrichment, soil protection and the provision of tree products.

An improved fallow can be a single- or mixed-species lot of low shrubs or ground cover or it could be a mixed-species stand that resembles dense natural regrowth. A mixed-species multistorey fallow might also resemble a carefully cultivated garden or a woodlot with regularly spaced trees. Fallows are usually impermanent, although they may be planned to evolve into contour vegetation strips in cropland (see section 4.2) or other, more permanent agroforestry systems.

If a fallow is to evolve into alley cropping or cropland with dispersed trees, this must be planned specifically. More attention to plant spacing and more labour-intensive planting methods will be needed when the fallow is established. If the fallow is to develop into another agroforestry practice, species choice and spacing should follow the guidelines for that particular practice. In addition, the woody plants that are to remain on site during future fallow cycles must be able to compete with other species in the fallow. This requirement tends to favour aggressive pioneer species which might become a problem at a later stage. Farmers may also use oversized seedlings or cuttings to give a 'head start' to the more permanent plants.

As with most of the practices described in this manual, improved fallows usually require protection from browsing and grazing animals. The degree of protection required depends on the number and type of free-roaming animals, herding practices and the vulnerability of the fallow vegetation to browsing, grazing and trampling. Most nitrogen-fixing leguminous plants are readily eaten by wild and domestic animals, especially when they are young, and trampling can be a serious problem, even for non-edible species. Many fields will have been protected when cropped, by fencing or other means, and this protection should continue through the fallow period.

Living fences may be established during the cropping period (see section 6.1) so that they are already in place during the fallow period. The living fence will have a better chance of becoming established when there are valuable crops in the field than later when the field is fallow. Such an approach requires advance planning and a high degree of motivation, which implies some previous demonstration of beneficial results. Initially, wire or thorn fences may be used to protect improved fallows. In many of the more densely settled areas of Africa, such as the highlands of Kenya and Rwanda, animals are well controlled, so that conditions are especially favourable for introducing woody plants into the fallow cycle.

SPECIES

While the main function of the fallow is to maintain or restore soil fertility and reduce erosion, some plants may be introduced primarily for their economic value. Species choice should not be confined exclusively to 'soil improvers', but rather plants with marketable products should also be considered if these are favoured by land owners or land users. Plants included in improved fallows should be compatible with future crops, free of any negative physical or chemical effects on the soil and not in competition with the crops to be planted later on the same site.

The degree of land pressure and the extent to which the local economy is commercialized will affect farmers' priorities concerning soil-fertility enhancement versus cash-crop production. In highly commercialized areas, farmers may prefer a cash-crop fallow of fuelwood, timber, fruit or high-protein fodder, with soil improvement as a by-product. For example, valuable trees could be established in association with leguminous cover crops for a net improvement in soil fertility, product diversity and overall economic benefit. In more diverse, subsistence-focussed systems, farmers may want soil improvement along with products for their own use, such as fibre, animal fodder, leafy vegetables, fuelwood and building poles.

Acacia mearnsii (black wattle), for instance, though less productive than other exotic species in terms of wood yield, produces tannin as a cash crop and has a well-documented record of restoring soil fertility for subsequent crop production. Other promising species include *Leucaena leucocephala* (where site conditions are favourable), *Sesbania, Gliricidia* and *Calliandra* species. These, in turn, could be mixed with valuable timber species, such as *Markhamia, Cedrela, Polyscias*

Degraded land in Niger planted with tree cuttings.

Improving Fallows with Trees

A SEQUENCE OF OPERATIONS THROUGH SEVERAL YEARS

NOTE: THE SEQUENCE SHOWN HERE HAS NOT BEEN CARRIED OUT IN THE FIELD. IT IS ONLY A SUGGESTION. MODIFICATIONS WILL BE NECESSARY TO FIT REAL NEEDS.

① THE FARMLAND IS LEFT FOR FALLOW.

② THE FOLLOWING RAINY SEASON TREE AND LEGUME COVER CROPS ARE PLANTED TO PRODUCE DENSE STAND.

③ TREES ARE GROWING, AND NATURAL VEGETATION ALSO INSTALLS ITSELF. SELECTIVE CUTTING FOR POLES AND FUEL WOOD CAN TAKE PLACE.

④ THE SOIL RECOVERS, AND THE LAND IS READY TO BE FARMED AGAIN. TREE LINES ARE MARKED, AND TREES BETWEEN THE LINES HARVESTED.

⑤ THE LAND IS CLEARED AND RE-PLANTED, BEING FARMED ON AN ALLEY-CROPPING SYSTEM, OR SIMPLY WITH TREES AND SHRUBS ON THE CONTOUR.

THE TIMBER TREES IN THE ALLEYS CAN BE LEFT TO GROW TALL, ONLY BEING PRUNED TO REDUCE SHADE.

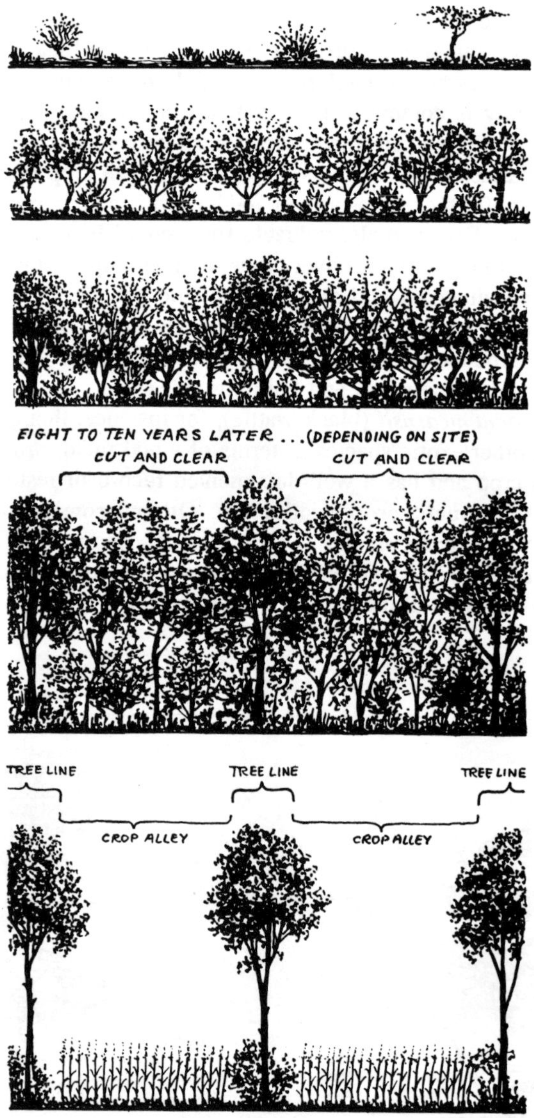

and, at higher elevations, *Podocarpus* species. Such timber trees could be left standing, either widely dispersed or in clumps, when the field is cleared later for cultivation.

Some soil-enriching dryland pulses are also excellent food sources. *Vigna subterranean* (Bambara groundnut), a native of Timbuctu at the edge of the Sahara, is popular across a wide area of dryland Africa, from Senegal to Kenya. Its seeds contain 14 to 25% protein, 6 to 7% fat and about 60% carbohydrate. Lablab beans (*Lablab purpureus*), which tolerate annual rainfall as low as 200 to 400 mm, are also widely consumed by people in the region; they contain 25% protein and 1.4% fat. Their leaves are eaten by goats, cattle, sheep and pigs. *Tylosema esculentum*, the marama bean from the Kalahari, is a dietary staple in Southern Africa. Its seeds, containing 30% protein and 36 to 43 % oil, are roasted or boiled, and the tuber is also eaten. *Cajanus cajan* (pigeon pea) is another useful and popular addition to a dryland fallow.

ESTABLISHMENT

Improved fallows can be established in a variety of ways, and at various stages of the fallow. Methods might include:

- direct seeding of clean-tilled, harvested plots
- selective cutting of bush, followed by enrichment planting with tall seedlings
- introducing tall seedlings and cuttings into poor-quality fallows on degraded land
- planting tree seedlings into closely spaced, deep planting holes or furrows within blocks of cleared cropland.

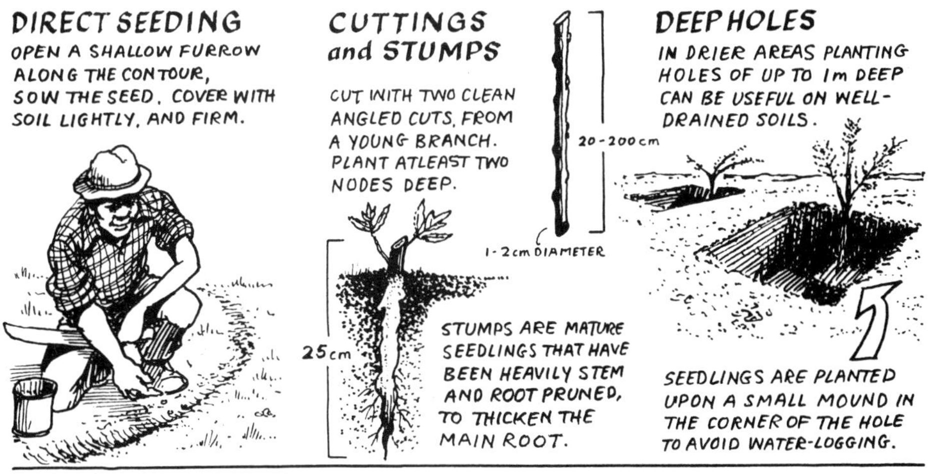

The exact techniques vary with previous land use, value of the fallow vegetation, condition of the land and expected duration of the fallow. Decisions on labour and cost allocation for establishing trees also depend on whether trees will be kept on cropland in future rotations.

MAINTENANCE

Once a fallow area is protected from animals and the vegetation is established, it requires little maintenance, limited to occasional weeding, pruning and harvesting of fruit, timber or other products. More labour is required for fallows composed of harvestable cash crops, as opposed to fallows primarily of pioneer species which need little care.

ANTICIPATED BENEFITS

While few formal improved-fallow experiments have been conducted with woody species, those which have been documented suggest that this practice has great promise for sustainable production in dryland Africa. Improved fallows, whether fully planted or selectively planted and managed along with natural regrowth, can serve both sustainability and production goals without interfering with cropping operations. With a minimum of labour, farmers can obtain edible leaves, fruit, animal and bee fodder, timber, fuelwood, fibre and craft wood from their fallow fields, while at the same time shortening the time required for recovery of soil fertility. Most of the labour required is concentrated in the infrequent task of land preparation for the next crop rotation.

Some woody legumes can add up to 300 kg/ha of nitrogen to the soil. Tejwani, in Mongi and Huxley (1979), reports on soil nutrients added by trees in a semi-arid region of India as follows:

Soil nutrients added by leaf litter (kg/ha/yr)

	N	P	K	Ca	Mg	Organic C
Natural forest	23.0	1.3	11.0	34.0	5.0	N.A.
Eucalyptus globulus	25.4	0.7	3.5	15.5	1.5	1163
Acacia mearnsii	19.5	0.4	2.9	3.0	0.7	516

While the benefits of fallow systems are usually measured in terms of improved crop yields, the more immediate economic benefits can be substantial. For example, a field planted exclusively to *Sesbania sesban* and left for 4 years can produce firewood and high-quality animal fodder, as well as soil nutrients.

The net addition of nitrogen to the soil at the beginning of the next cropping cycle is probably the best indicator of potential benefit to crop yields since crop-yield experiments over the whole cropping cycle are subject to several other variables, such as rainfall, pests, diseases, weeds, temperature and wind. A wider variety of trees can be planted in fallows than in more intimate tree-crop systems, such as alley cropping or dispersed trees in cropland. The range of species is not limited by considerations of competition with the main crop for water and light. For the same reason management requirements are also less stringent.

Fallows can produce wild foods, such as fruits or leafy vegetables, which are high in nutritional value and well liked by local communities. These plants can be planted or simply encouraged as they appear in fallow regrowth. Thus, fallows can serve as pools of genetic diversity favouring useful pioneer species. While this approach cannot substitute for creating reserves of fully developed, stable plant associations, it can preserve a wide variety of local plants for future domestication or simply for continued use in fallow-based systems. This may be particularly important for medicinal plants which are rapidly disappearing.

COMBINATION WITH OTHER TECHNIQUES

Woodlots and commercial tree-crop plots can be used in rotation with annual crops to improve the soil, depending on the species. For example, *Acacia mearnsii*, which is grown for tannin and wood in the Kenya highlands, is considered unsuitable for close intercropping but substantially increases soil fertility if used in fallow rotations.

BLACK WATTLE

Enriched fallows may be planted with regularly spaced trees, dispersed or in lines or clumps, which will remain in the cropland when the rest of the fallow growth is removed. In this way the fallow can lead into a more intimate mixture of trees and crops during the next crop rotation (see section 4.1). Once established, trees and shrubs can be integrated into small earthwork structures (section 5.1) or contour vegetation strips (section 4.2) or may be maintained to stabilize channels (section 5.2) or mark boundaries (section 6.2).

EXAMPLES FROM THE FIELD

Experimental systems of improved fallow in Africa have been largely confined to herbaceous and shrubby fodder legumes in short rotations in high-potential farmlands. However, several traditional practices and a few notable experiments have proven extremely effective in maintaining and restoring soil fertility in croplands.

While fully developed fallow systems with *Sesbania sesban* have not been well documented, there are impressive reports of soil improvement through intercropping with this species in Kenya's Siaya and Kakamega Districts. This tree's nitrogen-release mechanism, while compatible with intercropping, seems better suited to short-rotation

CROTALARIA

LIMA BEAN

fallow systems. Researchers have confirmed farmer's observations that most of the nitrogen derived from these trees is not cycled through leaf litter, but rather through the decomposition of the root system accompanied by shedding of nitrogen-fixing nodules when the trees are felled. Maize grown on sites formerly occupied by this species is particularly large and vigorous. Given the relatively short time to maturation and the value of leaf-fodder and fuelwood production, *Sesbania sesban* and other hardy *Sesbania* species warrant formal testing in short-rotation fallows. Farmers in the same region also report that *Striga* weed is controlled by long fallows. During the fallow period, the weed is suppressed by the other vegetation.

Farmers in West Cameroon plant *Tephrosia* seeds into fallowed land, then cut the plants after a few years to farm again. In the Ruvuma region of southern Tanzania, farmers plant *Crotalaria ochroleuca* (sunhemp) to suppress weeds, replenish soil fertility, especially nitrogen, and combat nematode infestation in vegetable crops during the following rotation. The sunhemp is interplanted, usually with maize, and rotated with other crops, usually vegetables. Experienced farmers have suggested that sunhemp be introduced into depleted croplands to restore fertility. The plant provides cattle feed and is reported to add up to 300 kg/ha of nitrogen to the soil, in addition to controlling weeds and insect pests. It thrives in subhumid and humid conditions.

Some projects aimed at introducing fodder species have established plants in enriched fallows. The fodder crops are planted together with the last crop raised in the field. *Phaseolus atropurpureus* (siratro), *Macroptilum lathyroides* (phasey bean), *Lablab purpureus* (lablab bean), *Stylosanthes hamata* var. *verano*, and the grasses *Andropogon gayanus* and *Cenchrus ciliaris* have been established successfully in savannah regions in fallows where animals are controlled.

SELECTED REFERENCES

All Africa Press Service (1986). *Tanzania: legume replacing inorganic fertilizers* Nairobi: All Africa Press Service.

ILCA (1986). *Panesa Newsletter* (Pasture Network for Eastern and Southern Africa). Addis Ababa: ILCA.

Mongi, H.O. and Huxley, P.A., eds. (1979). *Soils research in agroforestry: proceedings of an expert consultation.* Nairobi: ICRAF, 585pp.

Sokoine University of Agriculture (1986). *Research report on Marajea* (Crotalaria ochroleuca). Morogoro, Tanzania: Sokoine University of Agriculture.

CHAPTER FIVE

AGROFORESTRY WITH STRUCTURAL CONSERVATION MEASURES

5.1 Trees, Shrubs and Grasses on Small Earthwork Structures

DESCRIPTION

Trees, shrubs and grasses can be used with several types of small earthwork structures, such as microcatchments, contour ridges, contour furrows, infiltration ditches, infiltration galleries or barriers placed along contour lines. The use of agroforestry practice to stabilize these structures and/or make them more productive will be described in this section. Section 5.2 covers agroforestry practices used with larger, more permanent conservation structures, such as broad-base and bench terraces, while section 5.3 discusses agroforestry practices used with gully-stabilization and channel-control structures. Section 5.4 covers the use of microcatchments for rainwater harvesting, specifically to improve site conditions for agroforestry practices in dry and degraded areas.

Small earthwork structures intercept and slow down runoff water, which prevents both sheet and rill erosion and in some cases also conserves water for plant growth in croplands, pastures and degraded sites under rehabilitation. Once interrupted, the runoff water is either trapped and left to soak into the soil or to evaporate, or it is channelled off sideways so that it does not spill over the structures and create gullies on its way downslope. In addition to conserving soil and water, most small structures provide favourable planting niches for trees, shrubs and grasses.

Small earthwork structures are widely used in soil conservation, forestry and watershed management programmes throughout Africa. Each region and country seems to emphasize one or two techniques for general use. For example, many farmers in dry areas of Kenya use contour bunds, protected by cutoff drains. Berms and bunds, combined with microcatchments, are used in dry agropastoral and pastoral lands from Sudan to Niger for growing trees, grasses and crops. They have also been used successfully in the semi-arid districts of Turkana, Baringo and Kitui in Kenya. In Zaire and Rwanda, contour furrows and infiltration ditches are more prevalent.

Traditionally, the use of small earthwork structures in Africa was limited mainly to small-scale irrigation systems. Most other structural soil conservation techniques were introduced either during colonial times or in association with development assistance programmes. Under colonial regimes, soil conservation structures, such as large terraces, were often built by forced labour. As a result, local communities may have negative feelings about this type of project.

Small earthwork structures are usually cheaper and easier to build and maintain than terraces. Where conditions of soil and slope permit, smaller structures may be as effective as terraces in conserving soil and water. Under conditions of rapid sheet erosion, some small structures, such as the *fanya juu* trenches (Kiswahili for 'make above', the earth dug from the trench being thrown uphill to form a berm) in Kenya, may eventually form wide benches or terraces (see section 5.2).

The contribution of plants to the stability of small structures is far more important than in the case of terraces. Trees and shrubs can

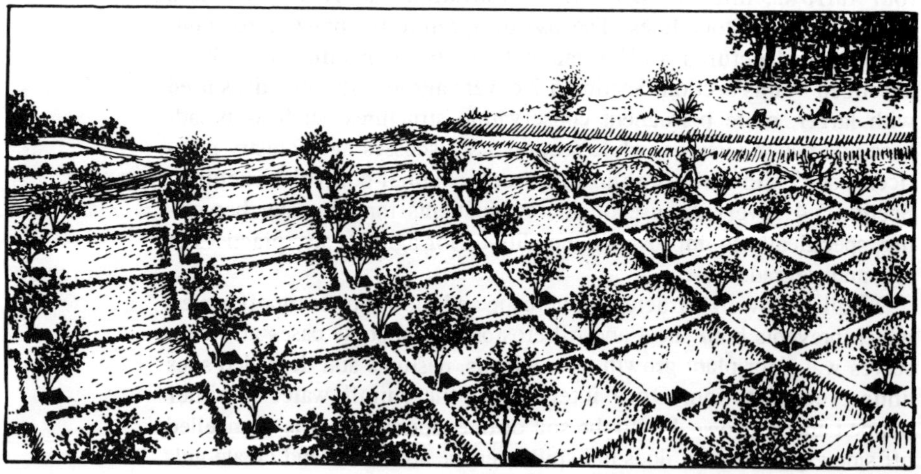

Small triangular microcatchments used to re-afforest a hillside.

protect the ridges and cut or filled slopes of newly constructed structures, making the difference between success and failure—especially in loose, sandy soils. Woody plants and grasses can also make 'lost' cropping space productive by using the surfaces of structures where other crops cannot be grown. With some soil types, the loose, deeply tilled surfaces of newly constructed ridges or furrows provide particularly favourable conditions for tree establishment. The combination of stored water and loose soil allows roots to sink deep before the next dry season. This is especially important in areas where soils normally tend to form a hard surface crust or where hard pans below the surface limit root penetration.

DESIGN

Since both earthwork structures and trees have a degree of permanence, land access and tenure must be carefully considered when designing an earthwork system. In order to be most effective, small structures, with or without trees, should be spread over an entire slope, as opposed to individual fields with some farmers participating and others not. The same is true for terraces: major problems occur if this practice is applied in scattered patches.

For structures built horizontally along the contour at zero grade, water must soak into the soil faster than it accumulates. Another approach is to drain water off sideways at non-erosive velocities. If the water drains down a slight grade, never over 2%, into a stable channel, it can continue down the slope without causing gullies or other forms of channel erosion.

If water is not drained off properly, runoff will be concentrated somewhere along the edge of a field or a series of fields. Gully forma-

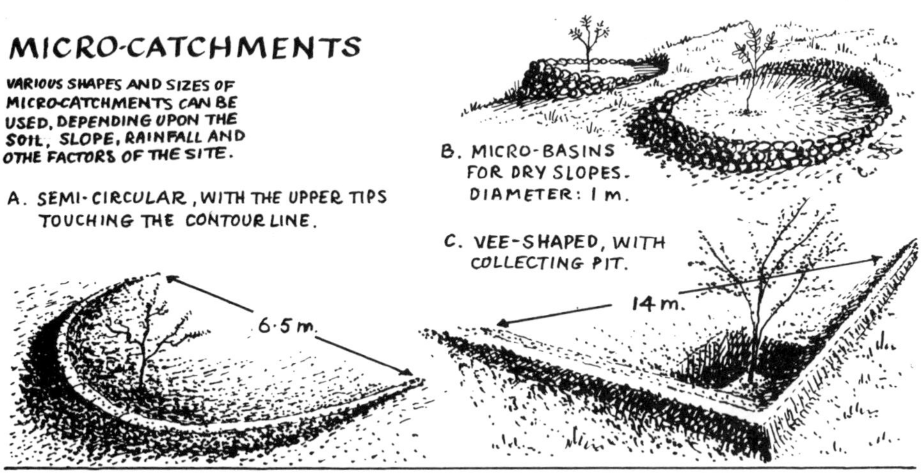

tion is inevitable unless measures are taken to control drainage. Otherwise, channel erosion and rapid runoff can cause major losses to land users downstream and to the community as a whole. Plots with earthworks also need protection from upslope drainage, often in the form of a cutoff ditch that collects and diverts water. Thus, grassed channels or other water-management techniques must be designed together with earthwork structures for soil and water conservation in order to prevent negative side effects. In both Malawi and Kenya, diversion ditches constructed on smallholder farms have sometimes concentrated runoff water, leading to severe gullying downstream.

A number of different small structures can be used on sloping cropland or grazing areas, depending upon the circumstances at the site. The choice and design of the structures will be described, followed by a discussion of planting options. The placement of trees on earthwork structures is shown in the illustrations.

EARTHWORK STRUCTURES

Proper spacing of ditches and ridges is extremely important. If these structures are too far apart they will be washed away or broken. If they are closer than necessary, then both labour and farmland are wasted. A few basic designs will be described here: Detailed methods of design and construction are given in specialized textbooks. It is important to remember that all formulae about the size and spacing of earthwork structures are site specific. Each site has its own characteristics that determine runoff and soil losses. Even the most precise calculations will only give approximate results. Field practitioners will always need to verify and adjust their calculations, based on experience with flows

CONTOUR RIDGES

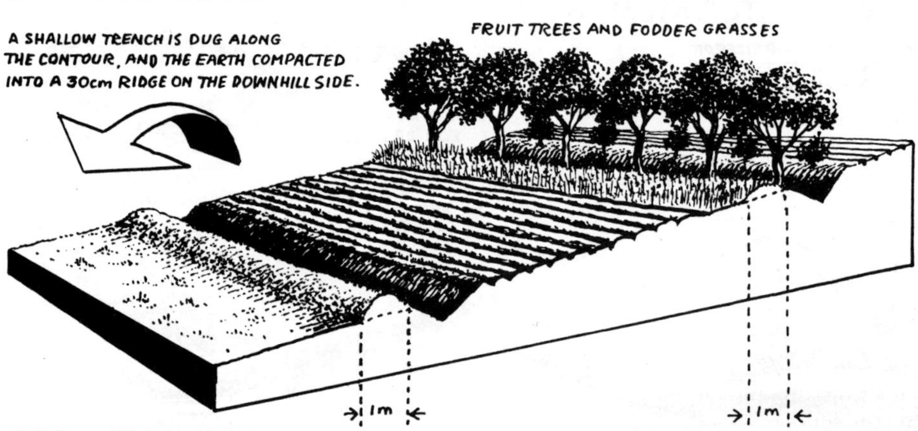

and erosion during heavy-rainfall periods. Since rainfall may vary dramatically from year to year, long-term local experience should always be taken into consideration.

Increasingly, detailed information is becoming available on rainfall and runoff in different environments and land-use conditions throughout Africa. National extension agencies for soil conservation, soil surveying and agriculture should be able to provide accurate local data to help in designing small earthwork structures. The following guidelines on type of structure, size and spacing may be helpful if more detailed or site-specific information is not available.

- *Microcatchments* can be built in dry locations mainly to trap and concentrate water in zones where crops or trees can profit from additional moisture. These are discussed in detail in section 5.3.
- *Contour ridges* are used in heavy soils that have low permeability. They consist of continuous ridges built with material excavated from nearby. The ridges act as small dams to keep water from running further downslope.
- *Contour furrows* are similar to ridges, but the water is trapped in trenches where it can either infiltrate or drain off sideways. One variation on this practice uses traction to dig individual furrows across fields along contour lines, using a simple harrow or plough. This process is repeated at regular intervals, dissecting the slope to create a series of parallel contour lines which increase the overall infiltration rate of the field.
- *Infiltration ditches* are level, medium-size trenches which collect runoff. They work best where the subsoil is more permeable than the topsoil. If the sides and the bottoms of these trenches allow water to infiltrate rapidly, they will absorb the surface flow completely and eliminate the risk of erosion.

CONTOUR FURROWS

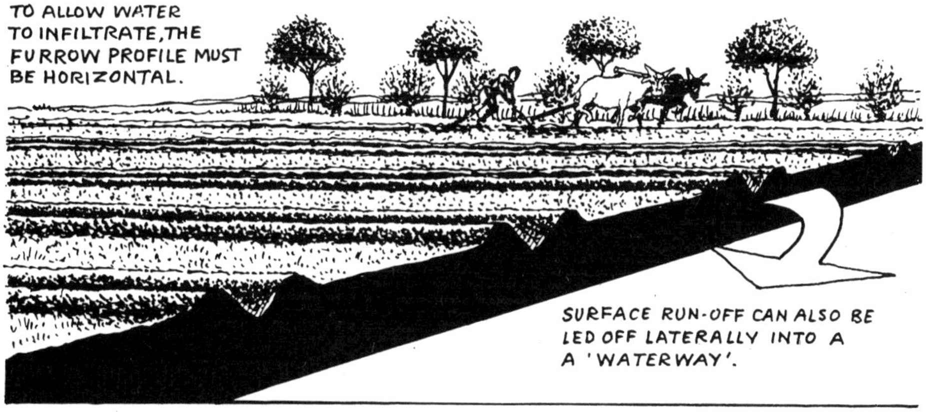

- *Infiltration galleries* can be used where soils are too impermeable for infiltration ditches. A contour trench is excavated and filled with loose rock, organic matter and soil to allow subsurface storage and to increase the infiltration of runoff water into the surrounding soil. Strips of grass and woody plants are usually placed upslope from the filled trench.

Any continuous, durable barrier placed on contour lines will at least slow down the flow of water long enough to break the erosive force of runoff. This will let soil particles and organic debris settle and form a band in which vegetation will establish itself naturally. Rocks can be placed in berms or 'trash lines' can be made by securing cut branches and other available material with stakes along the contour. The labour requirement is modest and enhanced over time by the natural deposit of additional material.

All small earthwork structures require maintenance. Wherever possible, land users should carry out maintenance and repair on shared or boundary lands cooperatively. Because structures on private land can be critical to an entire watershed, local agreement on the responsibility for maintenance is of the utmost importance. Each land user should be aware that a minimum amount of maintenance and repair will be required on a routine basis, and particularly after every heavy rainfall.

SPECIES

On small earthwork structures, the choice of what to plant, how to establish plants and how to arrange them is wider than it is for any other agroforestry practice described in this book. As in all agroforestry prac-

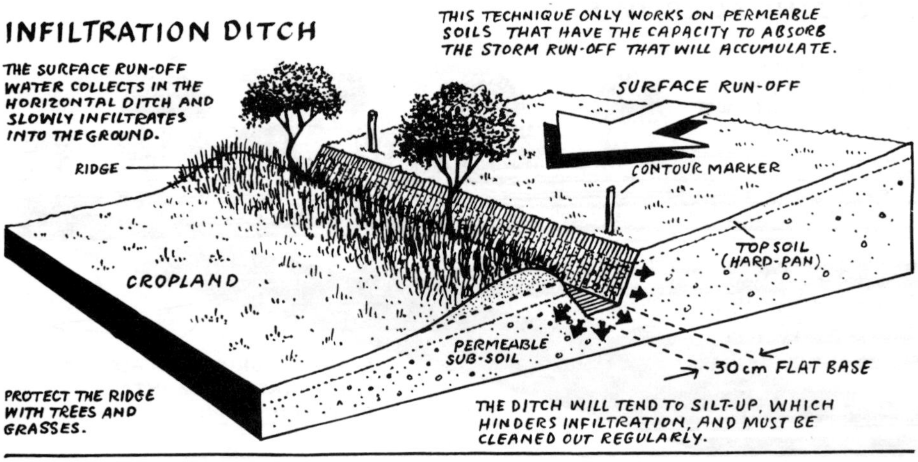

tices, the same basic rules apply for mixing trees and shrubs with crops. The woody plants selected should be:

- compatible with the cropping practice
- appropriate for the special soil and drainage conditions of the site, such as dry ridges, waterlogged depressions, compacted surrounding surfaces or loose, unconsolidated material
- useful for the provision of valuable products otherwise difficult to find or to grow in the vicinity.

In general the same sorts of plants and planting arrangements as those used on terraces work well on most earthwork structures, although some of these structures will not be as permanent as terraces. For this reason, valuable, slow-growing trees may not be a good choice for smaller structures, although the shorter life of small structures may not matter if the trees are able to withstand the disturbance of regular repair and reconstruction work.

Proper selection of species depends first of all on site conditions. The list of locally preferred species and the regional multipurpose tree species selection tables in Appendix I provide a first reference. Final species selection is a matter of farm management: Given the opportunity, people select those species most likely to meet their needs and priorities. In some cases, the emphasis may be on fruit trees, particularly those which do not create too much shade such as citrus, papaya or *Ziziphus* species. At higher elevations, Japanese plum or guava may be preferred. In some areas, species may be selected that produce particularly good-quality litter for mulching; in other cases, the main interest may be forage, fuelwood or bee fodder.

Some of the species commonly planted or left on soil bunds in the dry farmlands of India give an idea of the variety of tree types that can fill this niche. The favoured species range from *Salvadora persica* to *Dalbergia sissoo, Pongamia glabra, Prosopis cineraria, Acacia nilotica, Ziziphus* species and *Bauhinia purpurea.*

On steep slopes, effective erosion control through ditches and ridges requires close spacing. However, where horizontal spacing is less than 7 metres, even well-trimmed trees may adversely affect crop productivity. In such cases, small trees and shrubs may be preferable to large pole or fuelwood trees. Proper spacing and proper maintenance are also particularly important.

Pennisetum purpureum (Napier grass), *Setaria sphacelata* (Nandi grass) and sugar cane have been popular species for soil conservation structures in Kenya. *Acacia mearnsii* (black wattle) and banana are commonly grown along cutoff drains.

In many cases, earthwork structures favour rapid colonization by natural vegetation. Where this is true, it may be best simply to select and manage the most useful volunteer plants and gradually add other valuable species.

ANTICIPATED BENEFITS

Trees, shrubs and other permanent vegetation can substantially increase the effectiveness of small earthwork structures for erosion control. Permanent vegetation can also considerably reduce maintenance requirements. By stabilizing the soil, vegetation helps to retain the original line and grade of earthwork structures. A line of dense vegetation also discourages people and animals from crossing, and thus damaging, the structures.

Productive trees and shrubs occupy space along ditches or near the top of ridges which otherwise cannot be used to grow crops. The soil surface of earthwork structures usually cannot be cultivated without damaging the structures, so the soil becomes compacted over time. Farmers may be concerned that ditches or ridges take too much land out of production. If these spaces can produce wood, fruit, forage or other tree products, which for lack of land, water or protection cannot be grown elsewhere, then farmers may not view earthwork structures as 'lost' cropland. Trees and shrubs on small structures can also provide crops with light shade, shelter from wind and nutrients from increased leaf litter.

The dramatic effect of soil conservation measures on forage yield in an arid zone of Rajasthan, India, was averaged over a 9-year period from 1961 to 1970. Yields were measured in terms of kilograms of air-dried forage per hectare, as follows:

Treatment	Treated	Control	% Increase Due to Treatment
contour furrows	1566	213	635
contour bunds	1623	603	169
contour trenches	1321	490	170

Most of these improvements in yields were due to increased water availability (Ahuja *et al.*, 1973).

When incorporating trees and shrubs into these types of erosion-control structures, possible adverse effects should also be considered. For instance, trees can create too much shade over adjacent cropland. Proper species selection and careful management, including pruning, thinning, lopping and pollarding, can reduce this negative effect to tolerable levels.

The overall benefits of planting trees, shrubs and grasses on small earthwork structures are similar to the benefits of planting on large terraces. The main difference is the timing of costs and labour requirements. Small structures with woody plants are easier to construct, but must be maintained and rebuilt more often, while substantial terraces require more initial investment but less maintenance. Terraces also offer a more stable, permanent niche for trees.

COMBINATION WITH OTHER TECHNIQUES

Depending on the layout of farms and the location of livestock paths, woody plants on small earthwork structures may be used with living fences (see section 6.1), contour vegetation strips (section 4.2), dispersed trees in cropland (section 4.1) or pastures (Chapter 7). Combination with a special form of alley cropping (section 4.3) may also be beneficial. Some tree species planted on earthwork structures, when properly cut, can produce considerable amounts of leaf litter or mulch to protect the soil surface in the alleys between structures. This has yet to be proven in practice, but separate experiments conducted by IITA in Nigeria with alley cropping, mulching and small earthwork structures indicate that such a combination could be successful.

Semi-circular microcatchments in a pasture rehabilitation scheme.

EXAMPLES OF SUCCESSFUL ACTIVITIES

Several large-scale soil conservation schemes in the dry savannah zone of Burkina Faso and Niger have incorporated trees along newly built ridges. Where the trees were protected from grazing, they did well, but in most cases tree planting efforts were not successful. This experience emphasizes the importance of controlling livestock. Unfortunately, this problem is often addressed by considering physical means to control animals, rather than organizing the cooperation of owners and herders.

At sites in Burundi, Rwanda and Comoros, farmers have successfully planted ridges and other small structures with grasses (*Setaria splendida, S. sphacelata, Pennisetum purpureum, Tripsacum laxum,* and *Tripsacum vetiveria*) combined with woody plants (*Casuarina equisetifolia, Leucaena leucocephala, Sesbania* species, *Cajanus cajan* and *Grevillea robusta*). Results from the Nyabisindu project in a subhumid area of the Rwanda highlands indicate that trees should be planted on the upslope side of grass or bush barrier strips. Terraces form with time, stabilized by tree roots. *Grevillea robusta* is the most widely used species in this area.

In the subhumid western region of Cameroon, stabilization hedges have been planted on embankments, comprising single or double rows of shrubs. Following testing, this practice is now recommended to check runoff, trap soil, enrich the soil and diversify production.

Several projects in East Africa have combined trees, shrubs and grasses on small structures in croplands and as part of pasture-rehabilitation schemes. In some cases, such as grazing land-rehabilitation projects in Kenya's Kitui and Turkana Districts, field staff have modified earthwork structures to accommodate the special water-harvesting and protection requirements of trees.

Small earthwork dikes and strategically placed rocks were used in Senegal to reduce gullying in grazing areas. The dikes reduced runoff, trapped fine soil and increased water retention, resulting in better plant growth in adjacent areas of 200 to 400 square metres. However, in many places the dikes gave way to strong rushes of water. Further experiments showed that the dikes were partially effective but were too dispersed and should be located more densely in the drainage basin. Results suggested that live *Euphorbia* cuttings, with branches of a local plant laced in between, should be used to secure the dikes, rather than *Combretum* stakes, which were eaten by termites during the dry season.

In the *miombo* woodlands of northeastern Zambia, farmers who traditionally practice shifting cultivation have intensified cultivation in home gardens and selected permanent fields. In both sites, they have adopted a practice of cropping beans and cassava on beds of mounded soil over grass and leaf litter. The objective is to increase the fertility of the acid soils that characterize the region. On rolling hills and slightly sloping cropland, these mounds are built on the contour to control runoff and prevent erosion damage. In this case, tree leaves are used along with grass to make the mounds, rather than planting trees on the structures. Here, there is potential for a modified alley-cropping system, with hedges of mulch trees planted between the cropped mounds.

SELECTED REFERENCES

Douglas, M.G. (1988). *Integrating conservation into the farming system: land-use planning for smallholder farmers—concepts and procedures.* London: Commonwealth Secretariat, 130 pp.

FAO (1977). *Guidelines for watershed management.* FAO Conservation Guide No 1. Rome: FAO, 293 pp.

France, Ministère de Coopération (1979). *Conservation des sols au sud du Sahara*. Paris: CTFT, 296 pp.

Gupta, R. (1979). *Plants for environmental conservation.* Dehra Dun, India: Bishen Singh Mahendra Pal Singh, 247 pp.

Huxley, P.A., Rocheleau, D.E. and Wood, P.A. (1985). *Farm systems and agroforestry research in northern Zambia*. Nairobi: ICRAF, 76 pp.

Leblond, B. and Guerin, L. (1983). *Soil conservation: project design and implementation using labour-intensive techniques.* Geneva: ILO, 206 pp.

Wenner, C.G. (1981). *Soil conservation in Kenya: especially in small-scale farming in high-potential areas using labour-intensive methods.* Nairobi: Ministry of Agriculture and SIDA, 230 pp.

Tejwani, K.G. (1979). Soil fertility status maintenance and conservation for agroforestry systems on wasted land in India. In H.O. Mongi and P.A. Huxley, eds. *Soils research in agroforestry: proceedings of an expert consultation.* Nairobi: ICRAF, pp.141-74.

USDA (1954). *A manual on conservation of soil and water.* Agricultural Handbook No. 61. Washington, D.C: USDA, 350 pp.

Valet, S. (1985). *Action villageois bénévole de lutte anti-érosive dans les unités expérimentales de Thysse-Sonkorong en 1984.* Sine Saloum, Senegal: ISRA, 63 pp.

Young, A. (1986). *The potential of agroforestry for soil conservation. Part I: erosion control.* Working Paper No. 42. Nairobi: ICRAF, 68 pp.

5.2 Trees and Shrubs on Terraces

DESCRIPTION

This section describes the practice of combining large, permanent terraces with woody plants and annual crops. The term 'terrace' is popularly used to describe any soil conservation measure introduced

Hillside with bench terraces and a water way.

on slopes to change the natural course of runoff. Smaller earthworks such as ditches and ridges have been discussed in section 5.1. These are less costly to construct and generally less permanent than terraces. However, many of the principles and functions are the same.

Terraces are built mainly to conserve the soil and stabilize the slopes of steep land, while providing level areas for sustained cropping. The construction of terraces frequently increases yields and makes possible a wider variety of crops by improving soil-moisture conditions. Trees and shrubs may be used to stabilize a terrace and provide leaf mulch, shade and shelter from wind. In most cases, pre-existing trees and shrubs can be maintained on undisturbed parts of a slope or by adjusting terrace design and construction. Terraces may also improve site conditions for introducing valuable tree crops.

Terraces have been built traditionally in northern Togo, Mali, parts of Niger and in other countries throughout the world. They are now a common agricultural feature in many parts of Africa. To make a terrace, the slope of the land is cut and filled to make level or near-level steps. Large permanent terraces usually consist of a series of near vertical faces, called 'risers', composed of rock walls or steep banks of compacted earth. These are built to retain strips of level farmland in between in an otherwise steeply sloping area. Terraces with trees may also evolve over time by forming behind contour vegetation strips.

Two types of terraces will be described. These are broadbase ('conservation' or 'ridge') terraces, designed to remove or retain water on sloping land, and bench terraces, built to reduce the slope of land.

Broadbase terraces are actually a refinement of simple ridges, described in section 5.1. For a broadbase terrace, a slope is excavated to create near-level areas immediately upslope. These catch the surface runoff from the remaining uphill sloping land, allowing the water to accumulate in the excavated area. Water which is held back soaks the entire flattened area, increasing the amount of soil moisture available for crops. The slopes between the terraces may also be cultivated or protected with vegetation. The individual flat areas are generally not connected to each other, and resemble hillside contour ridges (section 5.1). Broadbase terraces are seldom used on slopes greater than 12% (Wenner, 1981).

Bench terraces are a series of more or less horizontal 'steps' cut into a slope. The excavated material is placed on the outside of the cut and an embankment is constructed so that the fill becomes part of the bench. The series of cuts and fills creates steps with level or near-level

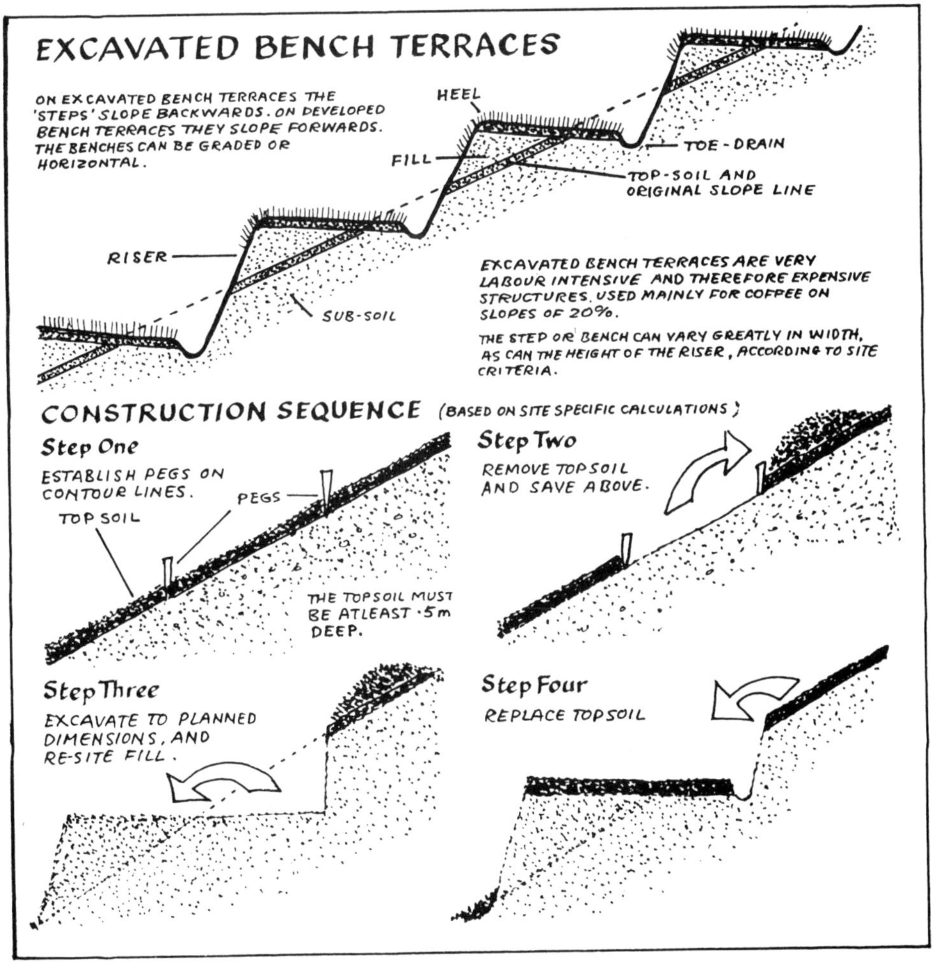

Construction of a 'Fanya Juu' *terrace.*

platforms. These cannot be constructed in shallow (less than 0.5 m) or highly variable soils, and mixing infertile subsoil with topsoil must be avoided. On steep slopes (25 to 55%) the benches become narrow ledges, especially suitable for fruit trees.

Bench terraces may also be formed more gradually, such as terraces which have evolved from *fanya juu* trenches in Kenya over a period of years. In this case, trenches are dug with ridges built upslope. The trenches should be as narrow as possible to minimize the amount of land taken out of cultivation. When covered with vegetation, the ridges catch eroding soil which helps to build them up further and create substantial flat areas behind them. The natural process of water carrying sediment is thus harnessed by carefully planted vegetation. This time-tested method uses a minimum of labour. Where soils are readily erodible, a fairly level bench may develop in 2 to 6 years. In Kenya's Nyeri District, *fanya juu* terraces reduced the hill slope by 4.5% after 2 years and by 7% after 4 years. However, under drier conditions, the necessary vegetation may take longer to develop, and may be damaged by droughts (Wenner, 1981).

DESIGN

The construction of new terraces or rehabilitation and improvement of old structures should begin with a sound structural design. Agroforestry practices must then be designed that are compatible with the structure and the crops to be grown. The species, spacing and management of woody plants on or between terraced fields may vary substantially according to the type and size of the structure.

Since terraces are permanent structures and involve substantial costs, labour and benefits, the long-term land tenure of an area to be terraced should be secure. As with many agroforestry practices, neighbouring land owners and users need to understand what a terracing project involves and need to agree on the level of cooperation and responsibility required, both in the short and long term. The plan should begin with a realistic assessment of whether terraces are really necessary and suitable for the existing production system and whether there are adequate skills, labour, capital and materials available locally to complete the structures and maintain them long afterward. In some situations, smaller structures or vegetation barriers may be less costly and more appropriate.

Trees can either be placed at the toe of the terrace riser or along its edge. In areas where soil moisture is scarce, tree roots find better growing conditions along the toe. In fact, the soil immediately behind the edge of the terrace is drier than anywhere else on the structure. In areas where rains and winds are heavy, trees planted along the terrace edge will protect crops and increase yields, although there may be a risk that trees will blow over and damage the terrace. A greater portion of the leaf litter will fall near the edge of the terrace if the trees are planted along the edge. Thus, in terms of site improvement and effect on crops, the best place for trees is at the edge of a terrace, whereas in terms of the tree's own requirements, the toe of the riser is best. Farmers will decide on the placement of trees on terraces taking into account the relative importance of tree services versus tree products.

Trees may help stabilize rock-wall terraces and the earth behind them, fastening themselves by sending roots into rock crevices deep

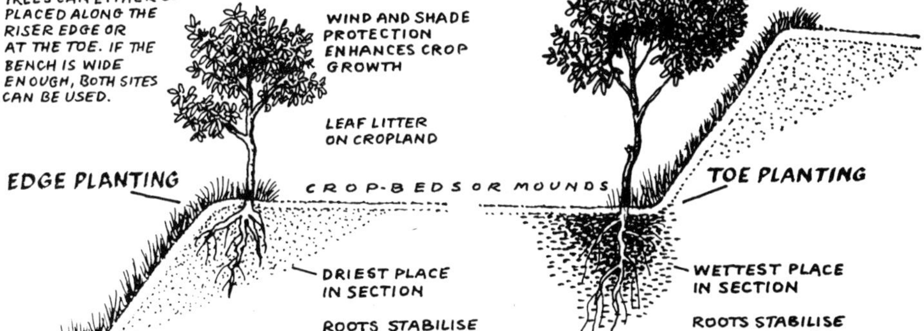

below the surface and acting as anchors to tie different soil layers together, thus reducing the chance of mass earth movements such as mud slides. However, tree roots may also dislodge rock walls.

Trees planted on any slope which is unstable due to a high level of soil moisture, may improve stability by absorbing some of the excessive water, thus reducing water pressure and helping to prevent soil slippage down the slope. However, in some cases, the forces of earth and water are of such magnitude that even the strongest tree roots will not be able to withstand them.

SPECIES

Many farming communities who have used terrace systems for generations have also practiced a kind of agroforestry by leaving and protecting trees of certain species in their fields. Trees such as *Grevillea robusta* may be scattered on terraced fields, spaced widely enough not to shade the crops but closely enough to contribute significant amounts of leaf litter, and thus organic matter, to the soil. In some cases the understorey may consist of small tree crops such as coffee.

Trees may also be mixed with staple food crops to produce fruit, fodder or wood. Fruit trees are usually planted along the edges of terraces. *Ziziphus, Olea* species (olives), and *Prunus amygdales* (almonds) are classic examples of tree crops planted on rock-wall terraces in many parts of Northern Africa and the Near East. More recently, other trees have been widely introduced on terraces in the region, particularly fruit trees such as *Persea americana* (avocado), *Carica papaya, Citrus* species and dwarf *Mangifera indica* (mango) and *Psidium guajava* (guava). Many farmers in Africa have also experimented with hedges of *Leucaena leucocephala* and similar trees on terrace risers to produce fodder and fuelwood on these previously underutilized sites.

ANTICIPATED BENEFITS

Farmers derive a number of benefits from combining woody plants and annual crops with terraces. For one thing, trees and shrubs complement the improvements associated with terrace structures—these may include stabilizing the slope, conserving topsoil and increasing water available below the soil surface. Trees planted on terraces may also affect temperature, wind, soil moisture, soil fertility and pests in ways that are beneficial to crops. Finally, broad, flat terraces allow farmers to practice methods of tillage—such as animal traction—which are not feasible on steep slopes. All these effects result in stable, or even increasing, crop yields that might otherwise decline over time due to land degradation.

Besides protecting and improving the existing cropping system, terraces provide new planting niches with favourable conditions for speciality crops or for establishing valuable trees. For example, farmers might plant fruit and nut trees along the toe of terrace risers on dry savannah hillslopes. The soil moisture conditions at the toe, along with the sheltering effect of a deep hole and the terrace riser, allow the successful establishment of tree crops that would otherwise be unable to survive under dry conditions.

The trees, shrubs and grasses planted on terraced cropland can protect and strengthen terrace structures. They can do this in two ways: by providing a surface cover of grass and leaf litter and by creating deep root networks in the subsoil. Trees and shrubs planted on the dry upper race risers may also shelter crops and improve the soil in this exposed, and often least productive, site on the terrace. Trees, shrubs and grasses can add to the diversity and value of products from the terrace, often using places that would otherwise be unproductive. Even where trees occupy favoured locations, such as the toe of risers, the value of the tree products usually far exceeds the value of other crops that could be grown on the same small area. Fruits, nuts, high-protein fodder and oil are all valuable tree products easily incorporated into terraces.

The combined effect on crop yields of agroforestry practices and soil-conservation structures—such as cut-off drains, trash lines and terraces—was calculated for one site in Kenya with an average annual rainfall of 1000 mm, 10 to 12% slope and soil composed of sandy loam. For farms where soil erosion was advanced, there was a dramatic increase in yield. Where measures were taken soon after conversion from pasture to cropland, before substantial soil erosion had occurred, the early high yields were maintained and even improved. According to the

Women tending a terraced garden edged with fruit and fodder trees.

farmers, maize yields increased by 62% and bean yields by 77% over yields before conservation structures were built. At least 50% of the increase in crop yields was attributed to the physical structures. Added to this was the value of new tree and shrub products (Hedfors, 1981; Wenner, 1981). For a detailed calculation of costs and benefits in such systems, see Sheng (1979).

COMBINATION WITH OTHER TECHNIQUES

Agroforestry practices to protect and stabilize open channels (see section 5.3) should be included as an integral part of terrace design. Other practices, such as dispersed trees on cropland (section 4.1), alley cropping (section 4.3) and multistorey gardens (section 4.4), may also be combined with terraces.

EXAMPLES FROM THE FIELD

In Burundi, Rwanda and the Comoros, gradually evolving terraces have been formed by planting lines of grasses with trees at 3- to 4-metre intervals. As part of a German Gesellschaft für technische Zusammenarbeit (GTZ) project in Rwanda, *Sesbania, Leucaena* and other small trees and shrubs were planted instead of grass between taller trees. In some areas, farmers are particularly interested in *Casuarina* trees for terrace risers because they fix nitrogen and do not give much shade.

In large areas of Kenya's Machakos District, farmers have planted *Macadamia* trees along the top of risers in terraced coffee fields. Banana and *Carica* (papaya) trees are planted in pits at the foot of the risers. This choice of species and placement takes advantage of differences in soil moisture and shade in different positions along the terrace profile. In the dry savannah zone in Machakos, many farmers plant *Citrus* species, papaya and bananas in pits along the foot of terrace risers in maize and bean fields. The terraces create a favorable microsite for these tree crops. Many farmers in the District have also begun to plant fodder trees and shrubs, mulch trees and a greater variety of fruit trees in this same niche.

Farmers have used terraces traditionally in North Africa and in scattered centres of hillslope agriculture elsewhere on the continent. Farmers in Africa now also use terraces widely for intensive, commercial vegetable gardens and cash-crop plantation on hillslopes, especially for coffee.

In some cases, such as in Machakos District, Kenya, terraces introduced in the past as a soil-conservation measure have been maintained and expanded by farmers to conserve water and improve crop yields. The use of trees on terraces is not yet widespread in most of Africa, although many farmers with terraced land have begun to plant trees along both the upper and lower edges of the structures. The popularity of tree planting on terraces in the arid and semi-arid regions of North Africa, the Middle East and India suggests a potential for wider adoption throughout Africa.

SELECTED REFERENCES

Anger, G., ed. (1985). *L'arbre et la haie dans l'exploitation paysanne.* Fiche Technique No. 3. Nyabisindu, Rwanda: Project Agropastoral de Nyabisindu, 126pp.

Hedfors, L. (1981). *Evaluation and economic appraisal of soil conservation in a pilot area.* Nairobi: Ministry of Agriculture, 24pp.

Pereira, H. (1983). *Land use and water resources.* London: Cambridge University Press, 246pp.

Sheng, T.C. (1979). Protection of cultivated slopes: terracing steep slopes in humid regions. In *Guidelines for watershed management.* FAO Conservation Guide 1. Rome: FAO, pp. 147-79.

Weber, F. and Hoskins, M. (1983b). *Soil conservation technical sheets.* Moscow, Idaho, USA: University of Idaho, 124pp.

Wenner, C.G. (1981). *Soil conservation in Kenya: especially in small-scale farming in high-potential areas using labour-intensive methods.* Nairobi: Ministry of Agriculture and SIDA, 230pp.

5.3 Protection and Stabilization of Waterways and Gullies

DESCRIPTION

Permanent vegetation, particularly trees and shrubs, can play a major role in stabilizing artificial waterways and gullies, as well as natural stream banks (see also section 6.4). Properly located in the channel section, woody vegetation helps decrease water velocity along the channel

Cut-off drain empties into a protected waterway.

edges and protects exposed soil, gravel or rocks from the erosive forces of flowing water. In smaller channels—even extremely steep ones such as gullies—trees, shrubs and grasses can reduce flow velocity across the entire channel. Once well established, these plants can completely stabilize small washouts and gullies and can complement physical erosion-control structures in larger channels. Trees, shrubs and herbaceous plants in such sites may also provide fuelwood, small poles, fodder, fruit, medicine, oilseeds or fibre.

Terraces, interception ditches and other erosion-control measures can rarely stabilize cultivated hillslopes without some complementary measures in drainage channels (see section 5.1). Unstable gullies and stream banks, fed by runoff upstream, frequently cut into otherwise well-managed cropland. To prevent this, upstream areas must be well managed and streams or gully channels must be stabilized. Sometimes the very structures used to stabilize hill slopes and control runoff on cropland or around houses may create new drainage problems downslope. For example, cutoff drains intercept surface runoff and channel it sideways at low velocity so as not to cause erosion. However, while the grade of these channels is not too steep—normally under 0.5%—, they all eventually come to an end. From this point, water is normally released to follow its natural course straight down the fall line. This artificial concentration of runoff can cause severe erosion unless the channels are protected.

Unstable stream banks and gully erosion are common throughout Africa, particularly in semi-arid farmlands. Efforts to stabilize gullies and stream banks most often rely on physical structures alone, while newly constructed waterways are often planted with one species of grass or remain unplanted. A few conservation projects use vegetation to stabilize channels, such as tree planting in gullies in Lesotho, and in many instances farmers establish gardens in stabilized gully sections.

The basic contribution that trees and shrubs can make in this context is physical. Through their root systems, the plants hold the soil and rocks lining the channels. The very presence of trunks, stems and strong root systems can also protect channel banks from the erosive force of water flow by increasing surface roughness: the added resistance helps slow down the velocity of moving water. This is especially important in the upper reaches of gullies where water moves quickly, since the erosive force of water is directly proportional to its velocity. Once the flow is slowed down, more water can seep into the soil. There, it recharges subsurface reservoirs or contributes to general soil moisture in the vicinity.

DESIGN

Three types of channel will be discussed: existing channels in need of preventive or repair treatment against erosion; gullies requiring stabilization and rehabilitation; and new, artificial or enlarged waterways that drain concentrated runoff from terraces, home compounds, roads or other structures. In some of these cases, design includes channel design and construction as well as the placement, establishment and management of vegetation at the site. However, only the planting aspects of design will be discussed here.

Although physical structures and the placement and spacing of vegetation vary for each of the three channel types, similar plant species

STREAM BANK PROTECTION

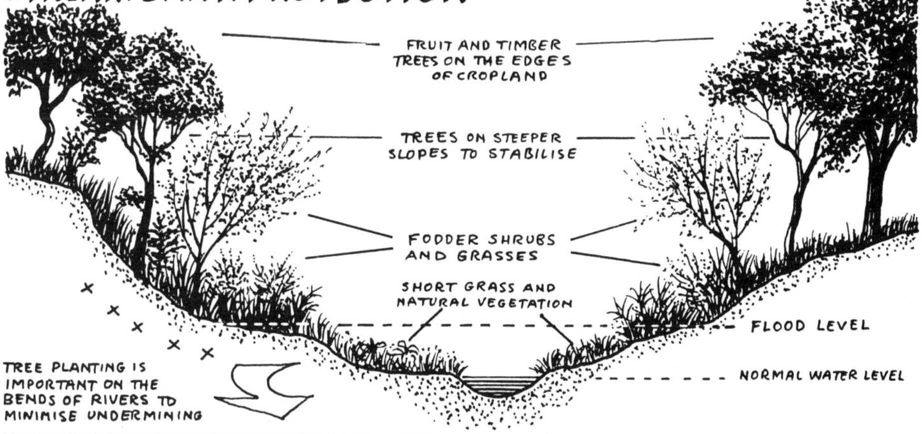

and management can be used. At the outset, it is important to determine whether channels may be controlled with vegetation or whether physical structures will also be necessary. The effectiveness of trees and shrubs in erosion control is limited where there is a constant and strong flow of water, such as in large, fairly permanent rivers. Existing channels may simply need to be protected with vegetation or they may be unable to carry the runoff and may need to be supplemented with other structures. This might include diversion of drainage to other, more stable channels.

One of the most difficult design problems is to estimate the level of peak flood that can be anticipated. The lack of information about the hydrologic characteristics of small watersheds in arid sites has led repeatedly to the design of inadequate structures, most of which wash out soon after construction. The level and strength of normal and flood flows must also be measured or estimated accurately enough the judge the size, type and placement of vegetation appropriate for a given part of the channel. This is true of stream channels, gullies and newly constructed waterways.

Large variations in rainfall between years can be misleading. For existing channels, there may be records of the level, speed and volume of previous flood peaks. In areas with few or no written records, the local people may remember details of previous floods. If they have lived in the area for several generations, they may have a considerable store of information. The older members of the community can often recall and point out the peak level reached at a particular point along a channel during 'the worst flood ever', as well as water levels reached in lesser storms occurring every year or every few years. It is best to talk with a number of people independently to obtain a reliable estimate.

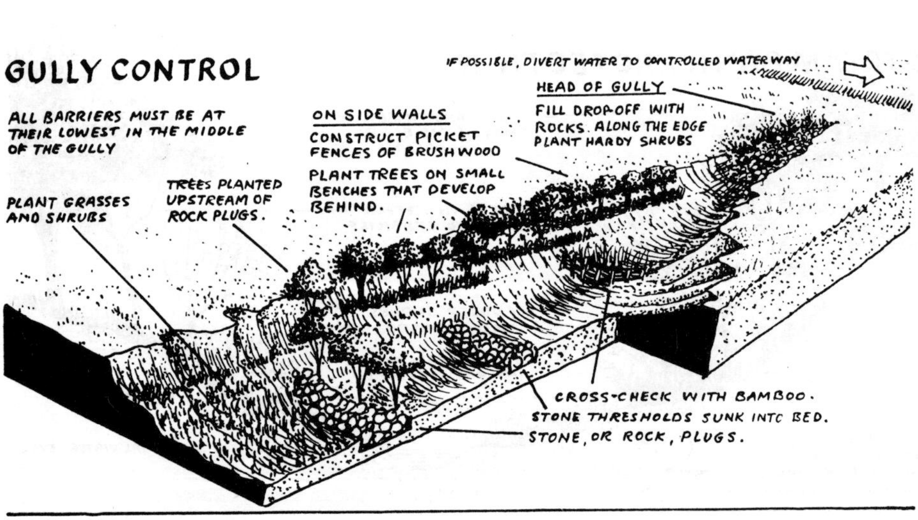

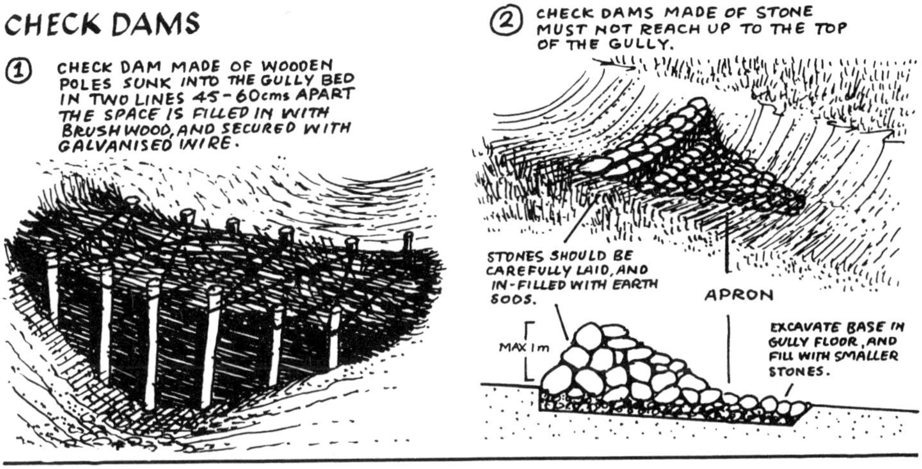

Information provided by the local community should be taken as an indicator to help derive a reasonable estimate, which is far superior to a wild guess. A field worker might decide to design a channel stabilization system based on the peak water level that appears likely to occur every 5 or 10 years.

A site visit, plus local information about flood frequencies and peak water levels, may be sufficient to plan stream-bank treatments. Farmers and field workers need to decide whether physical structures are necessary, where along the channel slope to plant trees, shrubs or grasses and when to plant them. Where a channel requires major water-control structures, it is important to take this process one step further—calculating the volume of water and the speed at which it moves during normal and flood flows. The methods for design and construction of such structures will not be treated here.

A gully might be treated as a dry stream channel and stabilized simply by planting short grass in the centre. If substantial drainage is diverted away from the gully into a nearby channel, that channel must be analysed to be sure that it can safely carry the diverted water flow. If it can, the gully may be filled in with rocks, earth and litter, and densely planted. First, trees, shrubs and tall grasses are planted on the upper banks and along check structures and then the built-up areas are planted as they fill in gradually through sedimentation. At each site, plants must have time to root before the first flood occurs.

For new waterways, the first task is to determine the size and shape of the channel. It is important to remember that different soils can tolerate different maximum water velocities before the channel erodes, scours or washes out. Much of the structural design depends on

whether or not the soil is to be covered with vegetation. The table presented below indicates how the maximum flow—and hence the ideal size of the waterway structure—will vary with and without vegetation cover under various conditions of soil, slope and volume of runoff. Once completed, newly constructed waterways can be treated like fairly stable gullies or streambanks, requiring vegetation and possibly small structures such as check dams. The permissible water-velocity ranges suggested in *Soil Conservation in Kenya* (Wenner, 1981) give an idea of the importance of soil type and plant cover in channels:

Maximum permissible velocities of water flow, based on soil texture and grass cover in channels (in metres per second)

Soil Type	Sparse Cover (dry areas)	Good Cover (subhumid areas, highlands, densely planted dry sites)
Silty sand	0.3	1.5
Sandy soil	0.8	2.1
Clay	1.5	2.4

These values are recommended only where flow occurs occasionally and the vegetation cover remains intact. If the cover is lost, then the maximum allowable velocities will be much lower. It usually takes up to 2 years to establish good grass cover; during this period, the channel is very sensitive to erosion.

Once established, vegetation can help control erosion levels in two ways. First, as the table shows, a good ground cover can prevent underlying soils from washing away when floods occur. Secondly, trees and shrubs along channel banks and across the channel floor (provided the water flow is not too deep) increase the roughness of the channel and thus reduce the velocity of water flow. However, a word of caution: if water velocity is reduced by trees and shrubs planted in a channel, then the channel must be much wider to allow the same volume of water to pass as when the water could flow freely.

If the site allows a slow flow of water over a wide channel, then introducing trees or shrubs along or in the channel bottom may be helpful. More often, however, stability in the centre of a channel can only be achieved by mechanical means, for instance using rocks or cement. In this situation, trees and shrubs are planted along the edges of the waterway. Their spacing and arrangement is extremely important. Trees planted along the edges of waterways can force running water into the narrow channel between them if they are not properly spaced and combined with grasses and low-growing shrubs. In such cases, the stream may cut deeper into the channel, finally undermining the trees.

SPECIES

In arid regions, there are usually many local woody species that thrive on the banks of streams that are dry most of the year. Such streams may only contain surface water during flash floods that last a few hours or less, but their dry channels contain more moisture than the surrounding areas because of subsurface storage. Several species of *Tamarix* and *Mitragyna* grow well in such sites. These are commonly found along small stream banks in both East and West Africa.

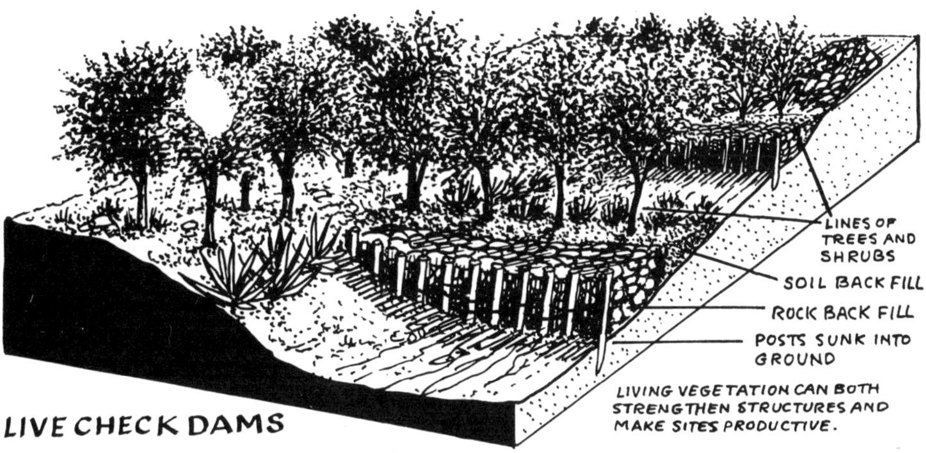

LIVE CHECK DAMS

Sesbania sesban and *S. bispinosa* also flourish along small channels and stream banks throughout the drylands of East Africa and in the Nile river valley. If immediate results are required, species that can be propagated from cuttings are especially effective, such as *Tamarix*. In more temperate areas, including the highlands of Ethiopia and Lesotho, members of the poplar family (*Populus*) are planted extensively along stream banks. Various species of bamboo (*Bambusa*), which are actually grasses, could also be used for this purpose.

The choice of species, spacing and planting design is quite different for gully control, especially if a gully is to be plugged and drainage directed to another channel. In this situation, soil and drainage conditions will change dramatically over time, so a succession of different plant combinations and arrangements is required, each adapted to a specific stage of gully rehabilitation.

MANAGEMENT

The management of vegetation planted along waterways and gullies often focusses on controlling access, rather than on managing the plants themselves. At these sites, there is usually enough water for growing plants, but it is important for plants to become well established before they are subjected to intense water flow. The most difficult task is usually to establish short grasses along the channel bottom. If sod was dug up during construction, it should be saved and used to line the channel.

Some plants may be established from direct seeding, while others, such as Napier grass, Kikuyu grass, *Commiphora, Euphorbia, Vitex, Mitragyna* and *Gliricidia sepium*, may be planted as stem cuttings. *Eucalyptus, Cassia*, Neem and *Melia* may be planted as root and stem cuttings, while other species are best established as oversized seedlings. Once planted, hardy species that are not affected by small mammals or insects (especially termites) should not require more than a good fence to maintain a stable, productive cover.

Whether plants along waterways are protected by a physical barrier or fence or by social agreement, people and animals must be prevented from trampling, harvesting or browsing the plants for at least two rainy seasons after they are established. If plants are established in a continuous strip along a stream or gully, then the design should include stable and well-marked crossings or access points for people and animals to reach the water source.

Once a waterway is stabilized and the plants along its edges are well established, then it is possible to begin the careful harvesting of plant products. This might involve controlled grazing, cutting fodder or harvesting fruit, fuelwood or timber from trees. Temporary waterways and gullies, which are dry most of the year, can be managed intensively.

Access and control are important aspects of the management of gullies or waterways because these sites often do not fit into clear-cut rules of individual or public ownership. The rights and responsibilities of local people to maintain waterways and use their products need to be clearly defined and widely recognized. Without well defined use rights, the people who maintain the channels cannot prevent others from destroying the structures and the vegetation through overgrazing or overharvesting. Rights and responsibilities may be assigned to specific families or to the larger community, perhaps organized in an informal association. Such a group could decide how to use the site and agree on the allocation of maintenance tasks and access to products.

A small group of users can manage a waterway on a daily basis, but after a heavy rainfall extensive repairs may be required before the next storm. A few individuals may not be willing to undertake such a large task in return for a few poles, fruit or fuelwood. In such a situation, a larger group of community members may be required to help with the work in light of the general benefits that the community derives from a stable drainage channel.

The management of a waterway requires a clear, effective agreement among community members and between the community and the relevant departments of government. One problem may be that government representatives are often reluctant to give control of a waterway to local groups, since this might imply indefinite continuation of use rights.

ANTICIPATED BENEFITS

The most important benefit of agroforestry practices along waterways is in terms of soil and water conservation. In areas where rock and cement are in short supply, trees and shrubs provide the only available conservation tool; in areas where other materials are available for channel construction, vegetation can strengthen and complement structures and make the site more productive. Valuable by-products from vegetation planted along waterways include fuelwood, poles, fodder for livestock or bees, gums, oilseeds and fibre. *Mitragyna* species, for example, are widely appreciated for making tool handles; bamboo from gullies is used for weaving mats and several species of *Eucalyptus* can be planted in gullies to produce poles and timber. The understorey may also provide some limited fodder grass or browse.

The careful use of sites along waterways may make a major contribution to the fuelwood and fodder needs of poor people with little or no land of their own. For instance, a series of stabilized gullies 2 kilometres long and 2.5 metres wide would cover an area of 5000 square metres. In 1 year, this area could produce enough fuelwood for 10 to 15 people, enough roofing poles for 10 houses and enough fodder for 5 cows.

COMBINATION WITH OTHER TECHNIQUES

Planting and stabilizing waterways, gullies and stream banks is an essential part of any effort to use physical structures and vegetation to terrace or otherwise stabilize cropland (see section 5.1). In addition, upland channels feed streams lower in the watershed: the management of upland channels complements programmes to revegetate floodplains and natural waterways downstream (section 6.4). Living fences (section 6.1) along one or both sides of a channel may also be useful for controlling access to the site by people and their animals.

EXAMPLES OF SUCCESSFUL ACTIVITIES

Several successful projects to stabilize riverbanks in dry areas near Agadez, Niger, have used shrubs and small trees. In another location in Niger, the banks of a medium-size, intermittent river have been at least partly stabilized with *Prosopis*. Several tree and shrub species have also been used with grasses and physical structures to stabilize gullies in dry savannah areas of Baringo District, Kenya.

Under conditions typical of many parts of Africa, bamboos are used to stabilize gullies in Haiti. This approach is also used successfully in the Philippines and Nepal. In the dry plains of southwestern India, *Prosopis* species are widely used to stabilize and protect gullies and, in the dry foothills of the Himalayas, *Vitex* and other woody species flourish in gullies in combination with rock structures. Many poor families gather fuelwood and fodder from such sites.

SELECTED REFERENCES

These texts tend to focus on physical structures to stabilize banks and gullies, but they also give suggestions on using shrubs and trees.

Dressler, J. and Neumann, I. (1982). Agriculture de couverture du sol. *Bulletin Agricole du Rwanda* Kigali.

FAO (1977). *Guidelines for watershed management.* FAO Conservation Guide No. 1. Rome: FAO, 293 pp.

France, Ministère de Coopération (1979). *Conservation des sols au sud du Sahara.* Paris: CTFT, 296 pp.

Gupta, R. (1979). *Plants for environmental conservation.* Dehra Dun, India: Bishen Singh Mahendra Pal Singh, 247 pp.

Leblond, B. and Guerin, L. (1983). *Soil conservation project for design and implementation using labour-intensive techniques.* Geneva: ILO, 206 pp.

Wenner, CG. (1980). *An outline of soil conservation in Kenya.* Nairobi: Ministry of Agriculture, Soil Conservation and Extension Unit, 57 pp.

Wenner,C.G. (1981). *Soil conservation in Kenya: especially in small-scale farming in high-potential areas using labour-intensive methods*. Nairobi: Ministry of Agriculture, 230 pp.

5.4 Microcatchments and Water Management

DESCRIPTION

In semi-arid and arid areas, rural people tend to rely more on livestock that can feed on natural vegetation—grasses, shrubs and trees—than on crop production, which is more severely limited by rainfall. Multi-purpose trees have supported people in these areas for generations. However, reduced access to land, recent droughts and other factors have led to the overuse, and in some cases disappearance, of existing trees and shrubs. In these areas, there is now a critical need to protect and plant trees for future use.

It is almost impossible to plant trees in semi-arid areas without some form of water management, unless the site is along a river or near an area that is seasonally flooded. Yet, little information is available about low-cost ways to improve water management for tree planting and agroforestry in dry areas. This section draws primarily on experience in Northern Kenya, with special thanks to Edmund Barrow, Forestry Advisor in Turkana District, Kenya, for his substantial contribution. The discussion will cover hand watering and various small-scale 'rain-water harvesting' techniques to establish trees, grasses and crops in areas with high temperatures and low and variable rainfall—from less than 200 to over 400 mm per year.

Collecting rainfall and runoff and diverting it to crops, livestock or household use is known as rainwater harvesting. This term usually refers to small-scale activities in which water is used in the same place where it is collected. Sometimes water is stored for livestock or household use, but this discussion focusses on situations where water is held in small structures on planting sites and allowed to infiltrate slowly into the soil. These structures are small and inexpensive and can fit into many situations and meet many different needs, such as fruit-tree production in a home compound, or the production of trees for fodder or fuelwood in rangelands.

Two rainwater-harvesting techniques are described here: triangular (or V-shaped) microcatchments and small semicircular microcatchments. Contour ridges and furrows may also be used for rainwater harvesting (see section 5.1). Like other techniques using earthworks, microcatchments are simply a means of collecting and guiding water to where it is needed—in this case, towards young trees. Much of the water is stored in the soil where it can be used by the trees until the next rains.

The size and layout of microcatchments should vary according to local conditions. Where rainfall is as high as 600 to 800 mm per year, a larger number of smaller structures should be used that will catch enough water but avoid flooding the tree or overflowing. In arid regions, the structure needs to cover a larger area in order to catch enough water to sustain a tree through the dry season. In either case, the exact size and number of structures should be determined by the volume of water that can be expected to accumulate on a given land area in the most intense rainstorm occurring over a 2- to 5-year period.

A triangular microcatchment is made of small, V-shaped bunds built by hand to direct rainfall and runoff downslope from a catchment area into a small basin at the lowest point in the structure. Trees, grasses and crops can be grown in the lower portion of the catchment where the water from the area upslope soaks into the soil, supplying adequate moisture for plant growth.

This system of using triangular bunds for establishing trees was developed in the Israeli desert, where the bunds are called *negarim.* They work best on sites with a modest slope, from 3 to 5%, where the soil allows for some runoff and infiltration. Microcatchments are versatile because they are small and can be built in sites where larger structures could not be used, such as around homes, schools or public places. They can also be used on rangelands, croplands, areas to be reforested and even along roads. When used for tree planting, their primary purpose is to provide enough water during the critical first and second year until young trees become well established.

Rainwater harvesting can also improve the growth and survival of naturally occurring young trees. It is often preferable, and usually easier, to encourage natural regrowth of certain local trees than to try to plant seeds or seedlings. For this purpose, a small semicircular bund is built by hand around the young tree to catch water from upslope and direct it towards the plant.

Naturally occurring trees in dry areas are often stunted by lack of water and continual browsing by goats and other animals. If the water available to young trees is increased, they will grow more quickly and will sooner be out of the reach of animals. For example, young *Acacia tortilis* trees appear to be stunted, thorny bushes. However, they grow quickly to their natural tree form if pruned, provided with adequate water and propped up to encourage straight growth out of the reach of goats. If sticks are used to prop up the leading branch, termite-resistant wood should be used.

DESIGN

The size and shape of a microcatchment or series of microcatchments depend on the intended use—for trees, crops or grasses—and on the soil type, rainfall pattern and slope of the site. The catchment area can range from 25 to over 100 square metres. Microcatchments can be arranged in many lines along the contour of a slope or scattered individually across the landscape. Where microcatchments are used across a hillside, they should be sited carefully so they do not create gullies, rills or overflowing.

A team of three people mark out a series of triangular microcatchments.

Triangular microcatchments

A triangular or v-shaped microcatchment that can hold 27.5 cubic metres (27,500 litres) of water would be appropriate for most semi-arid areas. This is large enough to hold over 200 mm of rainfall, so there is little chance of overflow under semi-arid conditions. This rather large capacity allows for differences in soil type, slope and rainfall conditions. It also provides for the storage of rainfall over a number of days. It is important to vary the spacing of these structures according to rainfall conditions. Approximate catchment sizes for a structure with a capacity of 27.5 cubic metres of water under different annual rainfall conditions are as follows:

Catchment Area (square metres)	Annual Rainfall (millimetres)
10	600
25	400
100	200

These dimensions can be adjusted to suit local conditions after testing and experimentation. The amount of water collected will vary according to rainfall as follows:

Rainfall (mm)	Water Collected (litres)
10	1,000
20	2,000
30	3,000
40	4,000
50	5,000
100	10,000

The type of soil in a particular site affects how much rainwater runs off the soil surface and how much soaks into the soil. Most of the rainfall should run downslope to the planting pit at the bottom corner of a microcatchment and then sink into the soil. Microcatchments do not function well in some soils. Sands are too permeable: all the rainfall soaks into the catchment area and none is collected. With heavy clays, virtually all the rainfall runs off the catchment area and is collected, but then it does not soak into the soil. Instead, the water stays at the surface where it is unavailable to the plant's deeper roots or it may waterlog the roots near the surface or even drown the young plant. In this situation, it may be appropriate to plant downslope from the microcatchment, where the tree can use the soil moisture but will not be waterlogged.

The best soil for a microcatchment is in between the two extremes—such as a sandy loam. The soil should also be fairly deep, so that water can be stored underground for plant growth during the dry season.

Triangular microcatchments can be laid out and built quite easily using hand labour. In northern Kenya, it took about 4 hours to build one triangular microcatchment of 10 square metres.

A low-cost method of constructing microcatchments has been suggested by the Forest Department of Turkana District, Kenya (1987). It requires:

- two pieces of wood 1.5 metres long, each with a notch cut in the same place near the end to tie a string, which serves as a stand for a line level
- a mason's line level
- a piece of strong string 14.5 metres long, knotted at the midpoint where the line level will hang
- a piece of string 20 metres long, also knotted at the midpoint.

If a series of microcatchments is to be built, then start laying them out at the highest point on the land and work downhill towards the lowest point. The line level is used to be sure that the two upper tips of the bunds are at the same elevation.

To use the line level, tie the 14.5-metre string to the notches on the wooden stands so there is 14 metres of string between them and the centre knot is 7 metres from each stand. When the line is pulled tight, the stands should be 14 meters apart. Place the stands on the slope roughly on the contour, with the string tight. Hang the line level on the string so that it is exactly at the midpoint.

If the stands are at the same elevation along the contour, the bubble will appear in the middle of the line level. If the bubble is not in the middle, move one of the stands up or down the slope. When the bubble is in the middle, the stands will be marking the positions of the two corners of the microcatchment. Put down stakes or large stones to mark these spots.

To lay out the bunds, one person should hold one end of the 20-metre piece of string at one stake (A). A second person should hold the other end of the string at stake B, while a third person takes the knot at the midpoint and walks downhill until the string becomes taut.

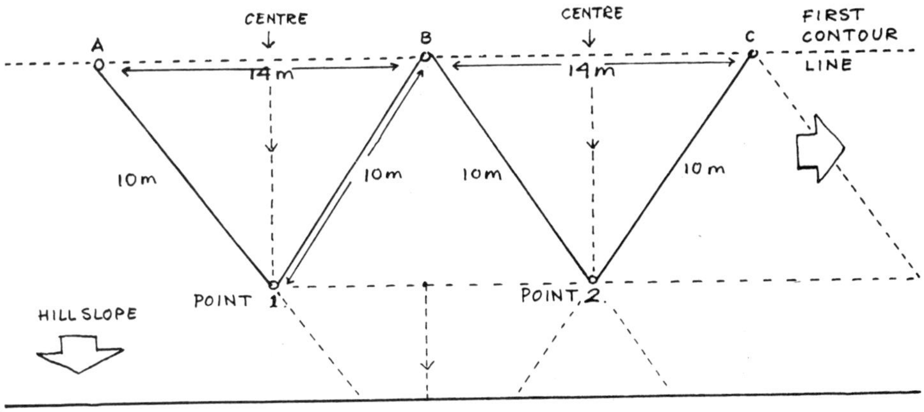

The string will now mark the two sides of the microcatchment. Put a stake or stone at the bottom where the knot is (point 1). This is the procedure for laying out a single V-shaped microcatchment.

Repeat this process to lay out a series of microcatchments. With points A and B marked using the 14-metre string, the person at point A moves to point B. The person at B then moves across to find a new point, C. Check that the bubble is in the middle of the line level and mark the spot with a stake or rock. Repeat the bund layout with the 20-metre string. This process can be repeated to create a line of catchments, all on the same contour with their uphill tips touching.

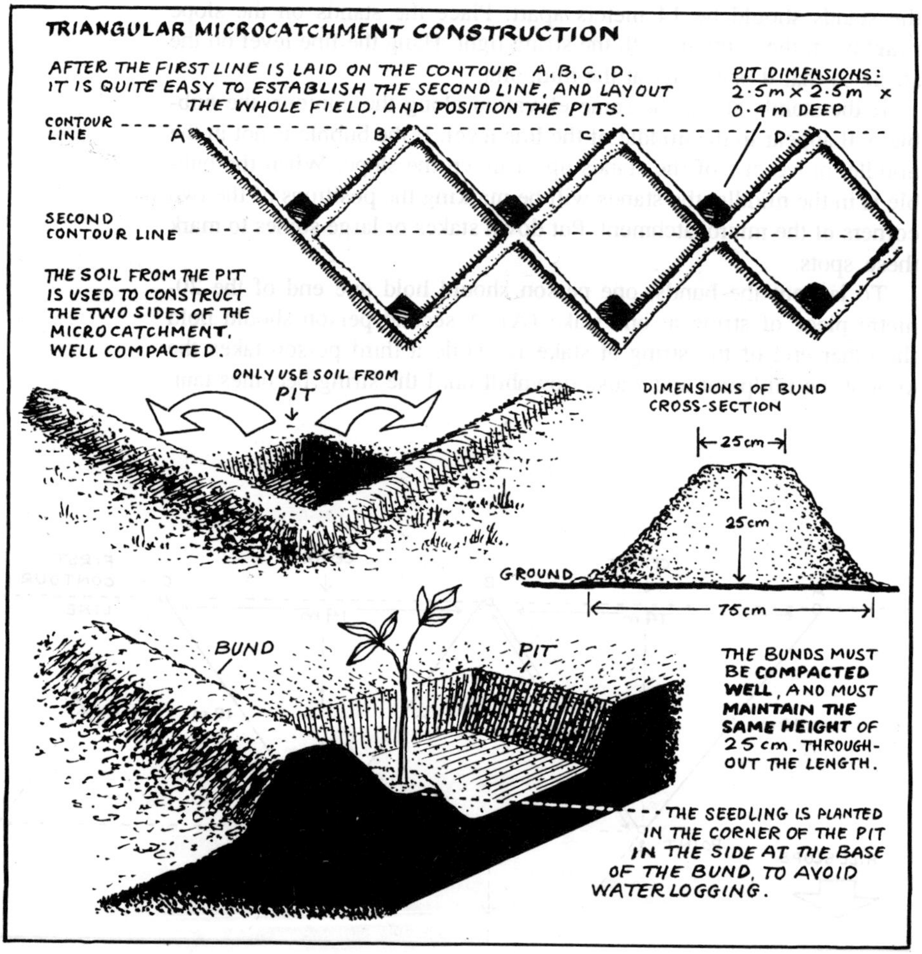

It is easy to lay out a whole field once the first line is laid out. Since points 1, 2, 3 etc. are all on the contour, use these points to lay out a second row. These points may not be the same distance apart, so move the stands up and down the sides of the catchments already laid out until the string is tight and the bubble is in the middle of the line level.

Soil taken from the pit dug at the bottom of each microcatchment to catch and store water is used to build the V-shaped bunds on either side. For a sandy loam soil and an annual rainfall of 200 to 400 mm, the pit and bunds should have the following dimensions:

- Pit: 2.5 x 2.5 metres wide x 0.4 metres deep; volume: 2.5 cubic metres
- Bunds: 10 metres long; 25 cm high; 75 cm wide at the base; 25 cm wide at the top; 2.49 cubic metres volume.

The bunds should be well compacted, for instance by stamping on them, and they should all be the same height. They should be built only with soil from the pit.

If no hand watering is planned, seedlings should be planted after the first major rainfall that completely fills the pit and covers the lowest part of the microcatchment with about 5 cm of water. This will be a rainfall of at least 25 mm. The seedlings should be planted when the water has completely soaked into the soil and the top layer has dried. In places where the soil bakes to a hard crust, some people prefer to dig the pits after the first rain when the soil is easier to work. If seedlings are to be hand watered, they can be planted at any time, with at least 20 litres of water added at the time of planting.

Seedlings should be planted at the base of the bund on the sides of the pit; in clay soils they can be placed downslope. If planted at the bottom of the pit, seedlings may become waterlogged and die. If planted in the catchment area or on top of the bunds, they will become too dry.

When planted, seedlings should be at least 20 cm tall. If no seedlings are available, 3 to 6 litres of goat manure can be dug into the pit, and after rain seedlings will germinate which can be thinned out. Grasses and leguminous herbaceous plants may also be seeded in the catchment along with trees. However, this is best done after the trees have become well established. The grasses should be managed so that they do not compete with the young trees for water or light.

Semicircular bunds

Semicircular bunds are made around existing trees or to promote growth of grass and leguminous ground cover. The tips should be on the contour, as with triangular microcatchments, and they should have a radius of about 3.25 metres. The tree should be 1 metre uphill from the lowest point in the semicircle. Use two pieces of wood connected

Semi-circular microcatchments allow Acacia tortilis *to grow and mature more quickly in dry areas.*

by a piece of string 6.5 metres long, knotted at the midpoint, and a line level. Using the line level and the same process as described for triangular bunds, locate the tips of the semicircle along the same contour (points A and B). Mark a spot 1 metre downslope from the tree (point C). Move the stands up or down the slope until the string is 3.25 metres uphill from the mark below the tree. Using the midpoint at the knot, draw a half-circle in the soil from point A to point B, going through point C. This will show where the bund should be built.

The bund should be constructed using soil scraped from the area inside the semicircle to a depth of about 10 cm. Be sure to avoid using the soil around the base of the tree or damaging roots near the surface. Build the bund 0.5 meters wide at the base, 0.25 meters wide at the top and 0.30 meters high. Remember to compact the soil well.

A bund constructed according to this design can store at least 1.65 cubic meters (1650 litres) of water after a rainfall. The actual amount will depend on the intensity and duration of the storm. More rainwater will be directed to the tree and stored in the soil than would occur naturally. In addition, if the tree is hand watered, the bund will ensure that the water soaks into the soil around the tree.

SPECIES

The choice of which trees to use in microcatchment plantings depends in large part on the preferences and needs of the local community. Since microcatchments are most often used in pastoral areas, trees which produce animal fodder may be the most desirable. Other locally important tree products might include fuelwood, wood for carving, medicine, fencing and foods. While exotic nitrogen-fixing species might be introduced to improve soil fertility, indigenous trees may be best adapted to semi-arid conditions and to the multiple uses important to local communities of herders and farmers.

In Kenya's Turkana District, several multipurpose trees and shrubs were favoured by participants in a Forestry Department training course: *Acacia tortilis, A. albida, Balanites pedicularis, B. aegyptiaca, Cordia sinensis, Dobera glabra, Delonix elata, Prosopis* species, *Salvadora persica, Tamarindus indica* and *Ziziphus mauritiana*, among others. Many Turkana have also shown interest in *Prosopis*, an exotic, since several species have been raised successfully in nurseries and transplanted into microcatchments in the area.

In trials conducted under UNESCO's Integrated Project for Arid Lands (IPAL) in Marsabit District, Kenya, goats preferred browsing *Acacia tortilis, Cordia sinensis* and *Cadaba farinosa*. All livestock were found to eat the flowers and fruits of *Acacia tortilis*.

The tables in Appendix I give an idea of the products from various tree species and their suitability for different climates and environmental conditions. The regional information can supplement observations of local species and uses, as discussed in Chapter 2.

Rainwater harvesting in microcatchments can also promote the growth of food crops, such as sorghum, cowpea and green gram, as well as pasture grasses. In some places people plant a mixture of grasses, shrubs and trees in microcatchments. Several national programmes in Kenya recommend planting indigenous perennial grasses mixed with some herbaceous legumes to avoid depleting soil nitrogen. Some of the best perennial grass species used in northern Kenya are *Cenchrus setigerus, Eragrostis superba* and *Chloris roxburghiana*. Strains of *Cenchrus ciliaris* and *Cynodon dactylon* have been successfully introduced in pastoral areas of Tanzania. Indigenous grass species planted in Turkana District include *Panicum coloratum, Chloris virgata, Eragrostis cilianensis* and *Sporobolus helvolus*.

When planting grasses with trees, it is usually best to allow the tree to become established alone during the first season. Grasses or food crops can be planted into microcatchments after the trees are growing vigorously so that in very dry years water consumption by these additional plants will not put stress on the trees.

BENEFITS

Rainwater harvesting in microcatchments is an inexpensive way to encourage plant growth in areas which are otherwise inhospitable to agroforestry. The construction of microcatchments is likely to be within the means of pastoral and agropastoral people who depend on the semi-arid environment for survival. Microcatchments may be used to encourage natural revegetation or to support planting efforts on degraded land. Rainwater-harvesting techniques are flexible. They can support trees in a variety of situations—from fodder trees in rangeland to fruit or fuelwood trees in farming areas. The small structures make agroforestry possible where water is not available from rivers or wells.

Even where water is available, microcatchments hold water near the trees so less water must be added to the planting site. This significantly reduces the time and effort required to carry water to trees. The use of microcatchments also maintains more water for tree roots at the end of the dry season when there may not be enough water to hand water trees as well as providing for human and livestock needs. In dry areas, even a small change in the use of water and labour may be decisive in determining the feasibility of agroforestry projects.

COMBINATION WITH OTHER TECHNIQUES

Rainwater harvesting can be used with many other agroforestry practices in order to enhance the survival of trees and shrubs. In areas where water is scarce, trees planted in cropland and pastures (see section 4.1 and Chapter 7), in windbreaks (section 6.3) and around houses and public places (section 6.6) can all benefit from microcatchments. Semi-circular bunds may also be used to encourage regeneration of selected species in improved fallows (section 4.5) and in pastures and rangelands (chapter 7). Where there is more than 500 mm of rainfall a year, catchments may not be necessary or may even result in the flooding of young trees, unless trees are placed on the ridge of the bund or just downslope of the area where water collects. In these situations, other types of small earthwork structure may be more effective.

EXAMPLES FROM THE FIELD

Semi-arid areas inhabited by pastoral people often receive less attention in soil and water conservation programmes than high-potential agricultural areas. In addition, large structures are usually emphasized, so there is little documentation on rainwater-harvesting projects. However, where microcatchments have been built, there has been considerable success. In northern Kenya, the Turkana District Forestry Department has demonstrated, 'time and time again that tree planting through the use of microcatchments is a viable means of successfully establishing trees in ASAL [arid and semi-arid land] areas [and] does not need extensive (and expensive) watering strategies' (Barrow, 1985).

In Kenya's Machakos District, farmers are using microcatchments to divert surface runoff to young trees from home compounds, livestock trails, grazing land and cropland. These small and simple structures have enabled farmers to grow a variety of indigenous and exotic trees in places where they would not otherwise survive. Combined with contour vegetation strips, microcatchments have been used successfully to reclaim gullied grazing land by promoting the growth of grasses, indigenous trees and introduced seedlings. In this area, farmers have limited access to permanent water so they must rely on managing rainfall. Their rainfall-harvesting efforts have an added benefit of diverting water from areas where runoff was causing soil erosion.

Microcatchments have also been used with agricultural crops and pasture grasses. A few handfuls of grass seed (*Eragrostis superba* and *Cenchrus ciliaris*) were spread before the rains in semicircular bunds on a test plot near Lake Baringo in Kenya. By the second season, grass was growing throughout the areas within the bunds. In Turkana, a single rainfall of 200 mm was enough to allow a family to harvest a crop of early-maturing sorghum sown immediately following the flooding of microcatchments. Substantial amounts of sorghum were harvested by many families following rains later in the season.

HAND WATERING

Microcatchments can be used to improve the use of natural rainfall, but hand watering may also be necessary. To be effective, water must be applied in a way that encourages trees to develop deep root systems that can reach the water table or moist soil far below the surface. Suggestions given here are based on the experience of the Forestry Department of Turkana District in semi-arid northern Kenya (1987).

Each seedling should be planted in a pit, with a diameter about the length of an arm; in this way, the limited water supply is confined near the plant. At the time of planting, the pit should be well soaked—with at least 20 litres of water. For the first 1 or 2 weeks, the seedling should be given 5 to 10 litres of water every 2 days. Before the roots are well developed, the young trees should be given about 10 litres of water every 4 days. Once the trees have started to grow again, about 2 months after planting, they should be watered once a month, with 20 to 40 litres of water.

This method has been developed from experience because too little water, frequently applied, results in a greater loss of water through evaporation. More importantly, frequent watering with small amounts produces a poorly formed root system, with roots growing towards the water source at the soil surface rather than reaching down to the water table or moist layers of subsoil.

The exact amount of water required depends on rainfall, temperature, soil type and slope. Sandy soils drain quickly so that seedlings need to be watered more frequently than on clayey soils that hold the water longer. In practice, the distance to water sources and the means available for transporting water also limit people's ability or willingness to water seedlings. For these reasons, the amounts and frequency of watering suggested here may be modified, but the principle of applying more water less frequently should be followed.

SELECTED REFERENCES

Barrow, E.G.C. (1985). *ASAL/TRDP quarterly report for the forestry sector, Turkana District.* Lodwar, Kenya: Forestry Department, 23 pp.

Barrow, E.G.C., ed. (1987). *Forestry manual for primary teachers in Turkana District.* Lodwar, Kenya: Forestry Department, 51 pp.

Finkel, M., ed. (1985). *Turkana water-harvesting manual.* Nairobi: prepared by Finkel and Finkel Ltd. for NORAD, 135 pp.

CHAPTER SIX

AGROFORESTRY FOR IN-BETWEEN PLACES

6.1 Living Fences

DESCRIPTION

Living fences (live fences) are a familiar feature throughout much of the African landscape. They appear on the densely populated hillsides of western Cameroon and in Rwanda and Burundi, marking small cultivated plots. In the dry rangelands of Northern Africa and the Sahel, living walls of vegetation form livestock enclosures and pathways to protect croplands and pasture from moving animals. Living fences are undoubtedly one of the most useful agroforestry techniques, as the need to control the movement of wild and domestic animals is a key element in most African land-use systems.

Some living fences are in fact converted plant fortifications, constructed originally to protect communities from aggressive neighbours and foreign invaders. In the Mandara mountains of northern Cameroon, the Kapsiki people constructed *Euphorbia* fortifications, some reinforced with stones, in the fourteenth century. These living fences enclosed fields and parks of valuable trees (*Acacia albida, Adansonia digitata, Ziziphus* species) and formed livestock corrals.

The elaborate defence system of the Midjiving in the Mandara foothills of Cameroon, built in the eighteenth century and still visible, illustrates the effectiveness of living fences. An outer line of *Commiphora africana* was extremely strong and provided fire resistance. A dense inner row of *Adenium obesum* reinforced it, making the system virtually impenetrable. These fences are reported to have stopped both mounted warriors and bullets at the beginning of colonization. The besieged economy was maintained by enclosed parks containing *Ficus gnaphalocarpa, Borassus aethiopum, Ziziphus* species, *Tamarindus indica, Vitex doniana, Celtis integrifolia, Acacia senegal* and *A. albida.*

The purpose of living fences may vary. Perhaps most often, people plant them to keep out domestic or wild animals. In other cases, they may demarcate areas where general access is discouraged, such as around a compound, house, cropland, fodder lot, garden or orchard. Living fences may also form livestock driveways and enclosures or separate fields which are in different rotations of crop or pasture management (paddocking).

The Midjiving used Adenium obesum *to control entry into their settlements in the Mandara foothills.*

In most living fences, the majority of plants are low, rarely over 2 metres tall, and the fence as a whole is dense and impenetrable. Trees or shrubs are planted close together in one or more rows. Living fences typically include plants which can be grown from large cuttings,

Euphorbia *hedges in a Luhyia compound in Western Kenya.*

such as *Euphorbia tirucalli.* As they grow, the plants are trimmed and, for some species, the branches are woven around the stems to form a thick, dense barrier. In other situations, full-size pollarded trees are used as living fence posts with dead branches, barbed wire or timber woven in between. Low, dense living fences may also contain individual trees at intervals which are allowed to grow to their full height.

Living fences differ from trees planted in rows primarily for soil conservation and improvement or for fodder or fuelwood production, as in alley cropping (section 4.3), or from trees planted in contour vegetation strips (section 4.2). The hedge structure in these cases is a by-product rather than the main purpose of the practice. Trees and shrubs on borderlines and along roads (sections 6.2 and 6.5) may not be intended primarily as barriers, although they can be designed to serve as living fences as well as their other functions.

DESIGN

Living fences should be considered to be permanent or semipermanent structures. They require maintenance and are likely to affect more than one land user. They can easily be removed, but the labour and costs invested in establishing them will be lost. It is better to locate them carefully where they can be of long-term benefit. In cases where a living fence is planned to demarcate a property line, all land owners and users affected should agree to its installation. They should also be aware of their rights (harvesting) and responsibilities (maintenance) regarding the fence. Different neighbours may have different priorities, for in-

A cattle-proof sisal hedge also provides material for basket making.

stance keeping insects or birds away from a field rather than enclosing large animals. It should be noted that beneficial, as well as troublesome, birds and insects may be attracted to living fences. People may also be accustomed to using paths which would be blocked off by a fence.

The basic design of a living fence is simple. Trees or shrubs are planted at 30- to 90-cm intervals in one or more rows, straight or in zigzags along the intended fence line. The requirement for density varies greatly with the purpose of the fence and woody species used. For example, a tight fence is required to keep out young goats, but if small animals are not a concern, a series of sturdy wooden stems may be adequate. Thorny trees, shrubs and vines which can twine through and over the fence are often included to make it more impenetrable. Since living fences are relatively permanent structures, they should not be planted so close to gardens or paths that they interfere with existing uses.

SPECIES

Plants which produce fruit, fuelwood, medicines, leafy vegetables or fodder can all be used as living fences with the correct site conditions and management. In the case of small gardens, fruit-tree orchards or tree nurseries, a living fence can also function as a windbreak (see section 6.3). Fruit trees used as a living fence can both protect and increase the productivity of small plots. Species such as *Moringa oleifera, Psidium guajava* and *Dobera glabra* are especially well suited to this role.

Wherever living fences border gardens or croplands, it is important to choose species that will not interfere with crop production when properly managed. Shade and root competition can generally be managed by thorough pruning. However, species with competitive rooting systems, which are aggressive, sprouting or self-seeding or which produce growth-inhibiting chemical substances or toxins should be avoided.

Living fences must also grow rapidly, at least in the first few years. Species that are easily propagated by large cuttings can be used to establish a fence quickly. Plants should be chosen that are easy to establish, especially from cuttings, that grow vigorously and that provide the expected functions and products. In general, sturdy, small trees or shrubs with multiple stems or low, dense branches are best. Plants with a good natural defence system, such as long thorns, spines or unpalatability, should be included. *Dovyalis caffra* (kei apple) is particularly popular for its spines. *Agave sisilana* (sisal) is also popular since it has sharp pointed leaf tips and produces light building poles and fibre for rope and basket making.

These natural defences may cause problems when fences are used near homes or public places. For instance, *Euphorbia* species have been widely used as living fences because they are easily established from cuttings, are inedible (except to camels) and form dense hedges. However, the traditional and very common *Euphorbia tirucalli* is now being replaced in areas of Kenya, especially around homes where children are present, because of its poisonous sap. Depending on site conditions and available stock, a variety of other woody species can be used, including *Ziziphus mauritiana, Z. mucronata, Commiphora africana, Erythrina abyssinica* and *Gliricidia sepium*. In Kenya's Siaya District, new fences of *Parkinsonia aculeata* are being planted inside the *Euphorbia*, which is cut down when the new trees have grown large

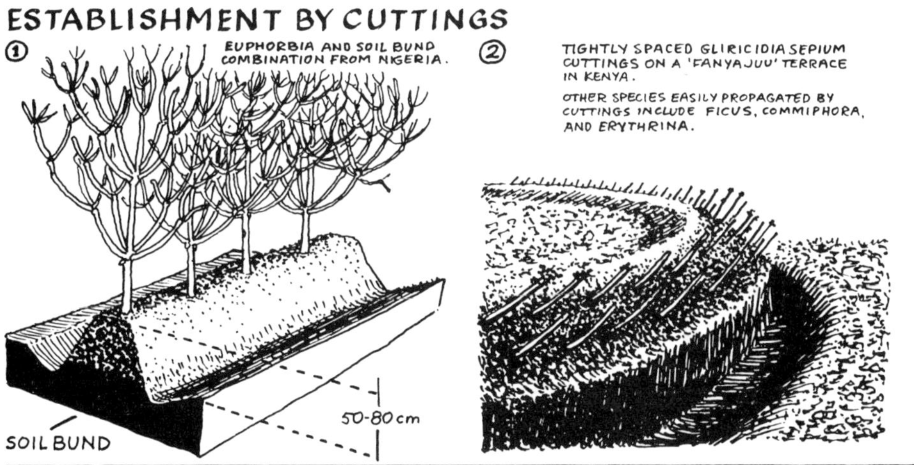

enough to be a substitute. See Table 2 in Appendix I for suggested species. Check Table 1 for special characteristics, such as thorniness, rapid growth and tolerance of pruning.

ESTABLISHMENT

Whenever possible, cuttings or seedlings should be used to establish a living fence quickly. The fence is then filled in with other species sown between the original plants. Alternatively, living fences can be started by direct seeding, especially in small garden plots where water is available and regular maintenance is possible.

During the first 2 to 5 years, a living fence requires careful maintenance. While the plants are small they must be protected from damage by people and animals. Weeds, fire or lack of sufficient water can also destroy a new fence. Not only should the fence be protected, but the area which it is intended to protect needs to be protected in another way or left unused until the fence becomes functional. Most fence plants also require training and pruning when they are young and supple to boost their vigour and give the appropriate shape. Neglect at this stage can be fatal.

Imported fencing material, such as metal posts, barbed wire or chicken wire, is often used to protect a young living fence temporarily. After the plants have grown enough to function as a fence themselves, this material can be moved to protect a new section of fence plantings. In practice, however, this approach may discourage proper maintenance of the plants, as people tend to regard the imported material as semipermanent. In addition, the cost of imported material must

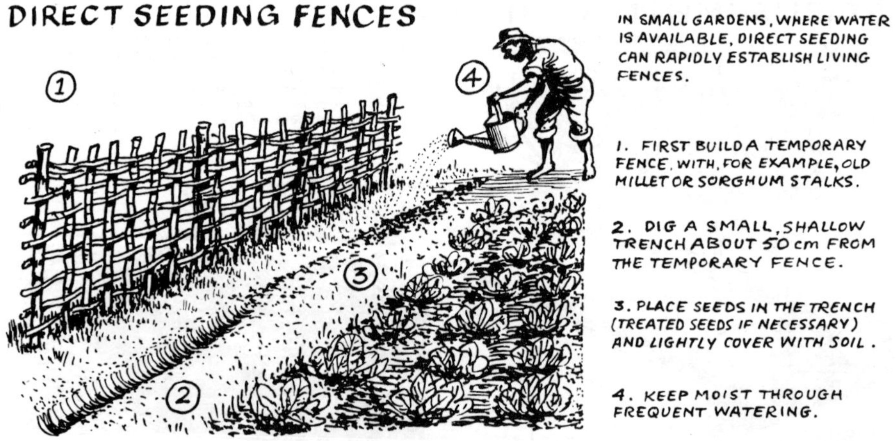

Small trenches prevent camels from browsing young trees.

usually be subsidized, which means that the success of the activity will depend on outside funding.

Throughout many of the pastoral areas of dryland Africa, people traditionally build fences with dead branches, usually cut from thorny trees such as *Ziziphus, Acacia* or *Balanites* or species considered as useless or undesirable weeds, such as *Lantana camara*. Such fences are also used to protect freshly established seedlings. The problem with this approach is that it requires large quantities of branches, putting additional stress on the remaining trees in the landscape. In some areas, the dead wood also attracts termites that may later attack seedlings in the living fence.

A small trench can also help to protect both the living fence and the enclosed area. This appears to work especially well to control camels. Microcatchments (see section 5.4) may also help the trees become established by increasing the available water and controlling localized erosion.

MANAGEMENT

Once well established, living fences require little management—but what they do require is critical. For some species, as soon as branches in the main fence structure have grown long enough, they can be cut or bent and woven in between the stems of the woody plants. This will work well to reinforce the fence, especially if the branches are thorny.

As the trees and shrubs grow, they should be trimmed on both sides and eventually on top so that they will not take up more space than necessary or cast too much shade on adjacent land. It is important to cut tree branches properly. Branches are often half-cut and then ripped off the tree leaving long scars in the bark. This damage can kill the trees or limit their growth.

Trees with short life spans may need to be replaced periodically. Individual trees can be replaced every year in a mature fence or periodically all the trees of one mature species can be harvested. The approach will depend on local priorities and the availability of labour.

ANTICIPATED BENEFITS

ADENIUM OBESUM

Properly designed and maintained living fences can play a vital role in resource management and agricultural development. If living fences have no other effect than keeping animals away from farm fields, gardens or young orchards, they have served their purpose well. Protection against free-roaming animals may make the difference between success and failure of tree-planting as well as crop production, while it may be difficult or disruptive to increase the vigilance or change the practices of herders, especially in places where livestock have traditionally been allowed to range freely. Modern metal fencing could serve the same function, but the costs are usually prohibitive and metal fencing is easily stolen.

Thus, living fences can make the difference between an agroforestry activity that is economically feasible from a local perspective and one that is not. In addition, living fences can improve the microclimate within small enclosures by reducing wind and soil-surface temperatures and can improve soil fertility and moisture by adding leaf litter. These benefits will depend upon the fence design and species used.

Like most technologies, living fences involve costs as well as benefits, often unequally distributed among different groups of people. A large-scale fencing programme may simply secure existing, possibly illegitimate, boundaries or may lead to new *de facto* rules of land use and access.

Beyond the farmer's practical problem of crop damage by animals looms the larger question of privatization of land. The protection of gardens, small irrigation schemes, cropland and small fodder plots often deprives someone else of access to land. Traditionally, such lands may have served as important common or free grazing areas, especially during the driest part of the year. While fencing land along an available water source may be an important new development for crop production, the same fence may block access to water for other people's animals. Such changes can cause severe disruption of traditional grazing patterns. For these reasons, all users, not just owners, of an area should be involved in the decision to introduce a living fence. This will help to ensure the fair distribution of costs and benefits.

COMBINATION WITH OTHER TECHNIQUES

Living fences may have a windbreak effect, particularly for protecting small gardens, fruit-tree orchards or tree nurseries. A living fence may also be planted on one side of a larger windbreak to limit access and protect the windbreak from browsing. In both situations, the design requirements for windbreaks (see section 6.3) must be reconciled with those for living fences. Similarly, living fences can be established along roads and paths (section 6.5) or property lines (section 6.2).

Living fences can be used to define and protect fodder lots, multistorey homegardens (section 4.4), block rotations in sylvopastoral systems (Chapter 7) or plots under alley cropping (section 4.3). They can be useful in limiting access to river banks, gullies or other areas that are subject to erosion due to human or animal traffic. In this way, living fences complement agroforestry efforts along erosion channels and waterways (sections 5.3 and 6.4) and are valuable additions to dryland irrigation schemes.

EXAMPLES FROM THE FIELD

Rural people throughout Africa have constructed living fences for generations. Recently, many projects have promoted living fences to protect gardens, tree nurseries and tree-planting sites. Even in very dry areas, living fences have been established successfully in conjunction with dune stabilization efforts, for example using *Euphorbia balsamifera* cuttings in Niger and Mauritania.

A living fence with fruit trees provides a useful windbreak for a kitchen garden.

On private land in Machakos District, the Kenya Forestry Research Institute (KEFRI) has experimented with a combination of traditional and exotic living-fence trees to protect efforts to rehabilitate grazing land. Self-help groups joined with individual property owners and KEFRI researchers to plant large *Commiphora* cuttings, 4 to 10 cm in diameter, along with *Cassia siamea* and *Prosopis juliflora* seedlings in a three-row fence line. They protected the new fence plants with deadwood and thorn branches.

These living fences will serve two purposes: to prevent the repeated harvesting of thorn branches for deadwood fences and to provide fuelwood for domestic use and pods for cattle and goat fodder. In addition, the *Commiphora africana* and *C.myrrha* trees produce gum, which is used for incense and could serve as a cash crop if the regional market were developed. Most importantly, the living fences will provide an effective protection for grazing land which the people in the community can afford. The self-help groups involved in the project have already received several requests to expand their efforts on other plots.

In Malawi, smallholder farmers plant *Agave sisilana* (sisal), *Caesalpinia decapetala, Commiphora africana* (kobo) and *Euphorbia tirucalli* (nkhadze) as living fences. The sisal fibres are used as string and the kobo produces gum and resin used as an insecticide.

It would be impossible to list all the successful introductions of living fences during the past 5 to 10 years in the countries of dryland Africa. Careful observation when travelling usually provides numerous examples close to home. Information provided by the local people or extension workers is usually sufficient as a basis for living-fence trials or extension efforts. Trials or extension should only be initiated if planting material is available, if people of the area really want a fence at the proposed site and if local community groups support the effort.

Successfully establishing and maintaining a living fence can make an important contribution to many different development efforts in rural areas of dryland Africa. The difficulties and constraints associated with tree planting, pasture improvement and gardening in many areas sometimes seem overwhelming—not enough water, too many animals and only limited interest and enthusiasm from local people and development personnel. At the same time, a living fence requires local discussion and planning plus a few years of growth before it becomes fully functional and development of the enclosed area can begin. Development workers are often simply in too much of a hurry.

SELECTED REFERENCES

Douglas, M.G. (1988). *Integrating conservation into the farming system: land-use planning for smallholder farmers—concepts and procedures.* London: Commonwealth Secretariat, 130 pp.

Jensens, J.W. (1986). *A landscape development plan for Kakayuni, Kenya: landscape planning and design with the application of agroforestry in the rehabilitation of eroded grazing lands.* M.Sc. thesis. Wageningen, The Netherlands: Department of Landscape Architecture, Agricultural University and Machakos, Kenya: Dryland Agroforestry Research Project, 165 pp.

Seignobos, C. (1980). Des fortifications végétales dans la zone Soudano-sahélienne (Tchad et Nord-Cameroun). *Cahiers ORSTOM* . Série Sciences Humaines. 17, 3-4: 191-222.

Weber, F. with Storey, C. (1986). *Reforestation in arid lands.* Arlington, Virginia, USA: VITA, 335 pp.

6.2 Trees and Shrubs on Borderlines and Boundaries

DESCRIPTION

Plantings along borderlines and boundaries adds multipurpose trees, shrubs and grasses to any space dividing properties or land uses. Although mainly used along the boundaries of farms, home compounds, herders' camps, pastures or scattered cropland plots, trees can also be planted along other kinds of boundary. The boundary site, by definition, implies a special situation with respect to land and tree tenure.

In many areas, the concept of planting trees and shrubs along farm or other property lines is well established. Even in areas where the practice is new, it is usually readily adopted. Borderline trees and shrubs do not require substantial labour for planting or maintenance, and farmers can use species that provide useful products from otherwise 'vacant' boundary areas. Conditions for boundary planting are especially favourable due to land-tenure reforms in many countries, where rural people who did not previously own separate plots now have legally defined property. The process of defining land tenure is affecting rural communities throughout the dryland regions of Africa; farmers now both need and have the legal right to mark their boundaries.

BALANITES AEGYPTIACA

Boundary plants may be widely or closely spaced, in single or multiple lines. The most common form of boundary planting consists of a single line of widely spaced trees and shrubs. In some areas, such as

Kisii District in Kenya, people plant long rectangular blocks of trees along boundaries. Trees planted along boundaries are distinct from living fences (see section 6.1), which may be placed along boundary lines but are intended primarily as physical barriers.

DESIGN

MORINGA OLEIFERA

If trees are to be planted along a property line affecting more than one land owner, the spacing between plants and their placement relative to the boundary will be important. The owners need to agree on how to use the boundary area, whether jointly or separately. If they undertake a shared planting directly on the boundary, then both owners need to agree on how to allocate the costs and benefits of the plants. Alternatively, they may make two separate, parallel lines of plants, placed at some distance from the boundary, but must take care that the two designs are compatible, with adequate space between. One must not shade out the other or invade the canopy or root zone of the other. Where only one owner wishes to plant near the boundary, this must not affect the crops, pastures or other property on the other side. Alternatively, the owner must agree to compensate for such effects by sharing products from the boundary area with the neighbour.

Establishing plantings on boundaries may confirm or otherwise affect disputed property lines. Field workers should be careful to discover all the hidden implications of introducing trees and shrubs along boundaries—including biases or reasons for enthusiastic support—before planting is undertaken.

SPECIES

ADANSONIA DIGITATA

Certain tree species have been used traditionally to identify property lines in different parts of Africa. In many instances, trees or shrubs that happened to grow along boundaries have simply been left and protected. In other areas, farmers have planted specific trees, such as baobab (*Adansonia digitata*), or clumps of grass to mark the dividing line between plots. For example, farmers in northern Nigeria have used *Andropogon gayanus*, while *Euphorbia* species are legally recognized in Kenya as boundary markers.

While many good agroforestry species may grow well on boundaries, local people often have definite preferences for certain species. Many trees that can be planted on boundaries provide welcome additional forage and fruit trees can also serve as boundary markers. Familiar fruit trees include *Mangifera indica* (mango), *Ficus* species (fig), *Carica papaya* (papaya) and *Citrus* species. Specific communities may also prefer *Tamarindus indica* (tamarind), *Moringa oleifera*, *Dobera glabra*

Tamarindus indica *used as a boundary marker with other fruit trees and a row of* Grevillea robusta.

or *Parkia biglobosa* (nere). In very dry areas, farmers may use *Acacia albida, Balanites* species, *Cordia sinensis* and *Ziziphus* species. Field staff associated with agricultural, forestry and conservation projects have often noted that local people are interested in planting trees and shrubs along boundaries, but lack appropriate planting stock.

In addition to traditional species, many farmers choose to plant fast-growing trees, such as *Eucalyptus*. Other farmers may plant *Casuarina*, which fixes nitrogen if properly inoculated. This species may be planted in lines, usually at intervals ranging from 2 to 10 metres, depending on the intended use of the wood and the desired size at harvest.

Table 1 in Appendix I shows some species suitable for planting on borderlines. Check Table 2 (column 27) for special characteristics that may be troublesome, such as invasiveness or shallow rooting systems.

It is sometimes useful to view boundary plantings in terms of numbers of trees in order to compare production potential with other practices, such as woodlots. For example, on a 1-hectare plot with a 400-metre border, a landowner can fit 80 timber trees at 5-metre intervals, 100 fruit trees at 4-metre intervals or 133 smaller trees at 3-metre intervals, which could be used to produce fuelwood, fodder and/or small poles. Depending on the species, it may be possible to combine the timber trees with an equal number of smaller trees and perhaps 20 fruit trees. If the trees are placed on a boundary and shared between neighbours, each owner will have half the total amount, which would constitute a substantial resource for most smallholders. The priority is to introduce trees, shrubs and other plants that are well adapted to local practices and site conditions, that can clearly define boundaries and that provide useful products or protect and conserve the site.

CASUARINA EQUISETIFOLIA

Boundary plantings must also be compatible with adjacent land use. If they include tall, straight timber trees, then they must not be allowed to grow beyond safe harvesting sizes. If trees are to stay on the site for a long time, then the owners must use species which will not become too large for the site or for local harvesting technology.

MANAGEMENT

The most important management problem when planting trees on borderlines and boundaries is the protection of newly established plants. Individual trees and shrubs can be protected with thorn branches or other small structures. In situations where plants are used on boundaries far from homes and croplands, protective measures need to be especially effective.

When choosing tree species and management practices, it is important to keep in mind the timing of plant growth and harvesting of plants and their products. Boundary sites, often far from the home compound, are difficult to control. Farmers may prefer fruit, nut or fodder trees with a predictable and discreet harvest season that does not conflict with other activities. This will allow the owners to supervise and reap the full harvest and to divide the harvest with neighbours if the border plants are shared.

Where tree roots may compete with crops for nutrients and water, the borderline roots should be pruned, either by deep tilling or digging a trench about 50 to 100 centimetres away from the tree line. Similarly, the tree parts above ground should be pruned to prevent competition with adjacent crops for light. Pruning back branches, coppicing or pollarding may also yield wood and fodder, but farmers need to take care that the particular tree species can tolerate these practices.

DOBERA GLABRA

Where timber or pole trees are part of the boundary planting, the design should include trees of different ages so that boundary markers are retained after the first harvest. The use of coppicing species, such as *Eucalyptus saligna* or *Gliricidia sepium*, can also substitute for staggered planting dates. A mixture of tall trees with small, coppicing pole trees will allow harvests of two or more products in different seasons or years.

Boundary plants should also be kept free of insect or animal pests that might affect neighbuoring crops, pastures or home compounds. Coffee and tea plantations and smallholder grain fields have suffered extensive damage from pests that inhabit or take refuge in boundary plants. Such pests may include flying insects, birds, small mammals or root nematodes and their incidence may vary significantly between closely related plant species and varieties. In many areas, pest damage to crops has made farmers and development workers wary of tree planting on or near croplands.

ANTICIPATED BENEFITS

Boundary plantings can stabilize a site, add organic matter and nutrients to the soil and furnish useful plant products. In many cases, the tree products obtained from boundary plantings can replace products previously gathered from woodland or scrubland further from the farmers' homes. This replacement can save household labour and reduce the over-harvesting of natural vegetation and degradation of related soil and water resources.

In one area of Machakos District, Kenya, it was estimated that 12.3% of the study site was composed of property lines and internal boundaries. Together with gully and stream borders (1.8%) and paths and roadsides (0.7%), the area of linear features available for planting was about 15% of the total site. Conservative estimates indicated that 50% of current local fuelwood needs and nearly 40% of fodder needs could be met by planting trees, shrubs and grasses on these strips of land (Rocheleau and van den Hoek, 1984).

COMBINATION WITH OTHER TECHNIQUES

A boundary planting can serve as a living fence (see section 6.1) if it is designed to do so. For example, farmers may plant individual trees 4 to 8 metres apart in the fence row and permit them to grow up above the fence. These are maintained along with the plants forming the fence. However, the purpose of these trees is mainly production and

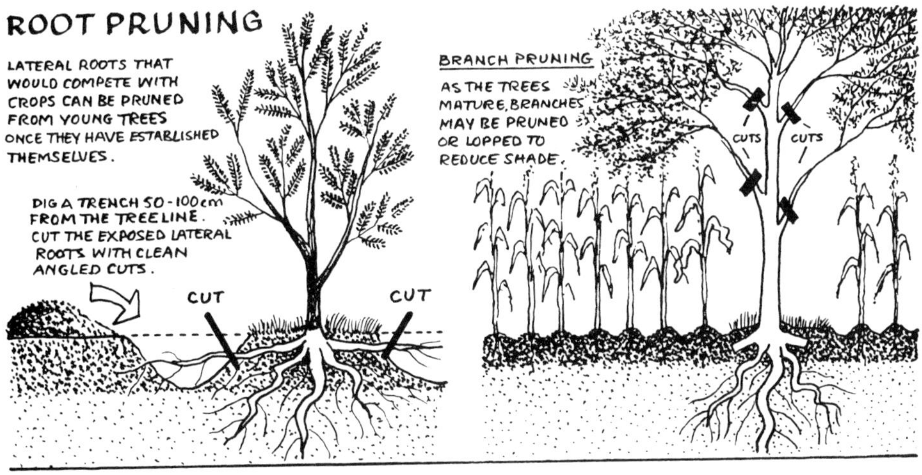

boundary demarcation, rather than fencing out animals or limiting access of people to the site.

Boundary plantings may complement open channel and waterway stabilization (sections 5.3 and 6.4) where drainageways coincide with boundary zones. Depending on the boundary alignments, boundary plantings may also be designed to function as windbreaks (section 6.3). Where long rectangular lots are planted along boundaries, the result is essentially a combined woodlot/boundary planting.

Boundary plantings represent a good opportunity to introduce more trees into a landscape where blocks of land are not available for planting trees. Ten or 20 trees planted along a boundary take less land out of production than if they are planted in a block, yet they may be equally, if not more, productive, depending on the product desired. It is important to choose the right species and to make sure that the people involved are convinced of the benefits of the planting, both for production and for boundary definition. Neighbours and community members must also have a firm agreement to protect and respect the trees and tree products in these shared, and sometimes vulnerable, sites.

EXAMPLES FROM THE FIELD

Many different types of project have supplemented or introduced boundary planting. Perhaps the most systematic effort is being carried out in association with several communal forestry/conservation projects in Rwanda, using a wide range of species according to local wishes and needs.

In Egypt, *Casuarina* species have been widely planted along boundaries, often in close association with grain crops, with careful management of the tree-crop interface. Farmers have learned to combat nematode infestation through choice of tree species, especially *Casuarina glauca*. They also control tree encroachment onto cropland by cutting trenches between crops and border plantings and allowing sheep to browse the new root shoots along the trench.

In Kisii District, Kenya, foresters distributed *Eucalyptus* species and *Cupressus lusitanica* seed and planting information to smallholder farmers over 30 years ago. The result was a proliferation of boundary line plantings and long, rectangular boundary lots nested into a dense patchwork of small farm plots. These now constitute the main source of fuelwood and building poles throughout the district.

In Kenya, as in many other countries, land surveyors often use trees as boundary markers. When land surveyors arrived to mark farmers' property at a KEFRI study site in Machakos District, the farmers and project staff agreed to use *Commiphora* and *Euphorbia* cuttings as the legally recognized boundary markers and then filled in boundary line with *Prosopis juliflora* and *Cassia siamea* seedlings. This cooperation

Eucalyptus saligna *used as boundary markers in a living fence.*

with government land surveyors resulted in the establishment and demonstration of live fences and boundary markers on several farms.

SELECTED REFERENCES

Anger, G., ed. (1985). *L'arbre et la haie dans l'exploitation paysanne.* Fiche Technique No. 3. Nyabisindu, Rwanda: Project Agropastoral de Nyabisindu, 126 pp.

Fortmann, L. and Bruce, J., eds. (1988). *Whose trees?* Boulder, Colorado, USA: Westview Press, 341 pp.

Fortmann, L. and Riddell, J. (1985). *Trees and tenure: an annotated bibliography for agroforesters and others.* Nairobi: ICRAF; Madison, Wisconsin, USA: Land Tenure Center, 135 pp.

National Academy of Sciences (1984). *Casuarinas: nitrogen-fixing trees for adverse sites.* Innovations in Tropical Reforestation Series. Washington, DC: National Academy Press, 118 pp.

Raintree, J., ed. (1987). *Land, trees and tenure: proceedings of an international workshop on tenure issues in agroforestry.* Nairobi: ICRAF; Madison, Wisconsin, USA: Land Tenure Centre, 412 pp.

Rocheleau, D. and van den Hoek, A. (1984). *The application of ecosystems and landscape analysis in agroforestry diagnosis and design.* Working Paper No. 11. Nairobi: ICRAF, 92 pp.

6.3 Windbreaks

DESCRIPTION

Windbreaks are strips of trees and/or shrubs planted to protect fields, homes, canals or other areas from wind and blowing soil or sand. Large-scale, wide strips or blocks of trees planted for this purpose are often called shelterbelts. Windbreaks are planted for many reasons: to reduce soil erosion, to improve the microclimate for growing crops and to shelter people and livestock. They can also serve other functions, such as fencing and boundary demarcation. Where wind is a major cause of soil erosion and moisture loss in dry areas, windbreaks can increase and sustain crop productivity.

When properly designed and maintained, windbreaks reduce the speed of the wind and thus its ability to carry and deposit soil and sand. They also improve growing conditions by decreasing water evaporation from soil and plants and can be used to reduce evaporation from water surfaces, such as irrigation ponds, canals or streams. In addition, windbreaks can provide a wide range of useful products, from poles and fuelwood to fruit, fodder, fiber and mulch.

Windbreaks made of vegetation usually consist of multistorey strips of trees and shrubs planted in one or more rows. Grasses and herbaceous plants are often planted at the base of the trees to prevent the wind from scouring the soil. Windbreaks are placed on the upwind side

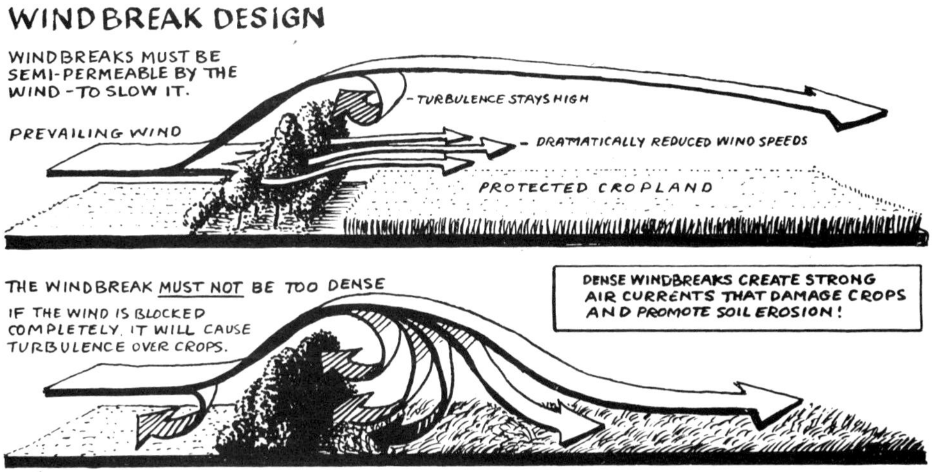

of the land to be protected and are most effective when oriented perpendicular (at 90 degrees) to the prevailing wind direction. The exact orientation also depends on the other roles which the windbreak may serve, as well as on property lines and topography. While size may vary, it is common in dry areas of Africa to plant windbreaks in lengths of 100 metres or more, with a mature height of around 10 metres.

Living fences and hedgerows can protect small sites, such as home gardens and nurseries, from wind. They may be planted on roadsides, boundaries or floodplains, but specifically designed to slow down the wind. Thus windbreaks differ from boundary plantings (section 6.2) and living fences (section 6.1) in terms of their orientation, which should face the wind. They should have multiple stories and be semi-permeable, letting some wind through. Very dense windbreaks may do more harm than good since they will tend to create strong air currents that will scour the soil on their upwind side and damage crops downwind. Gaps in very dense tree rows will channel the wind, actually increasing wind speed and contributing to soil erosion and crop damage on the downwind side.

Experiments are being conducted in Tanzania using dispersed trees in cropland to increase 'surface roughness' and thus decrease wind speeds, rather than planting distinct windbreaks. This approach avoids creating new, high-speed wind currents, but little is known about its effectiveness at present.

Throughout Africa, farmers use windbreaks to protect crops, water sources, soils and settlements on plains and gently rolling farmlands. Hedgerows of *Euphorbia tirucalli* (finger Euphorbia) protect maize fields and settlements in the dry savannahs of Tanzania and Kenya. Tall rows of *Casuarina* line thousands of canals and irrigated fields in Egypt. In Chad and Niger, multispecies shelterbelts protect wide expanses of cropland from desertification. The practice is not new, but agroforestry techniques can be applied to design windbreaks which will provide additional products and services to farmers.

DESIGN

The protective and productive benefits of windbreaks at a given site should be weighed against the costs before proceeding with detailed plans and planting. Aside from the direct cost of labour and planting material, windbreaks take land out of crop production and may compete with crops for water, light and nutrients. Increased crop yields, soil improvement and by-products must be sufficient to cover these costs and produce a net benefit. Farmers may also consider other constraints as they weigh their investment in this effort.

The greater the potential benefits, the better the chance of cooperation among land users over large expanses of land and across property lines. If individuals or families controlling separate plots are unwilling to cooperate with their neighbours, it will not be possible to develop optimally aligned large-scale shelterbelts. It is still possible to create windbreaks on individual plots if the holdings are large enough or if people are willing to plant a larger number of small structures. However, smaller windbreaks are less effective in protecting soil and water resources. Before a large-scale programme is undertaken, information on wind and other climatic factors should be gathered and analyzed for the area.

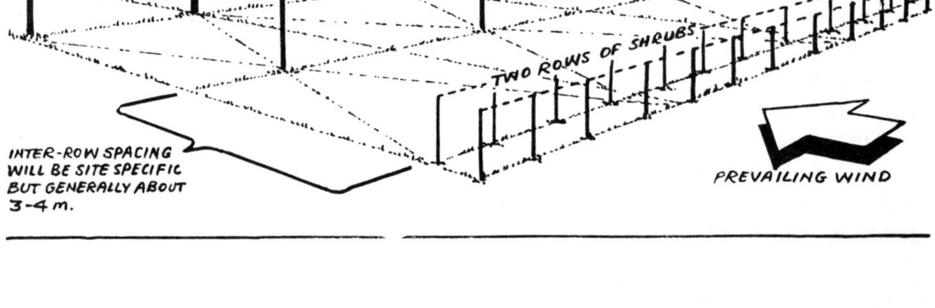

Where windbreaks span several individual plots or occupy shared land, planning should include an agreement concerning who benefits from any resulting by-products, such as fuelwood, poles, forage or fruits. This must be agreed before the trees are established, if possible even before the species choice and spacing is finalized. The same agreement should define responsibility for maintenance and harvest requirements. Products should be harvested without reducing the windbreak's effectiveness in protecting the crops and soil. People must know exactly *what* they can harvest and *how, how much* and *when.*

The orientation of the windbreak is crucial. In areas where the direction of prevailing winds changes from season to season, the local people can decide when is the most important period to provide protection. If soil erosion during the fallow period is the primary concern, then trees should be planted so that they are upwind of the cropland at this time. If, on the other hand, protecting crops is the primary need, then the windbreak should be placed so that it is upwind of the cropland during the growing season.

Where property lines, roads, buildings or natural features do not limit spacing, the distance between windbreaks is determined by the height of the tallest trees. A properly designed windbreak can protect a field at least 10 times as long as the height of the tallest trees, so, if the trees are 10 metres tall, crops up to 100 metres downwind will be protected. However, protection diminishes with distance away from the windbreak. Remember that it will take several years before the trees reach the optimum height to protect the full area. Remember, too, that a more permeable windbreak will shelter a longer stretch of cropland than a dense windbreak and also that windbreaks are most effective if repeated in a pattern of long strips throughout the landscape.

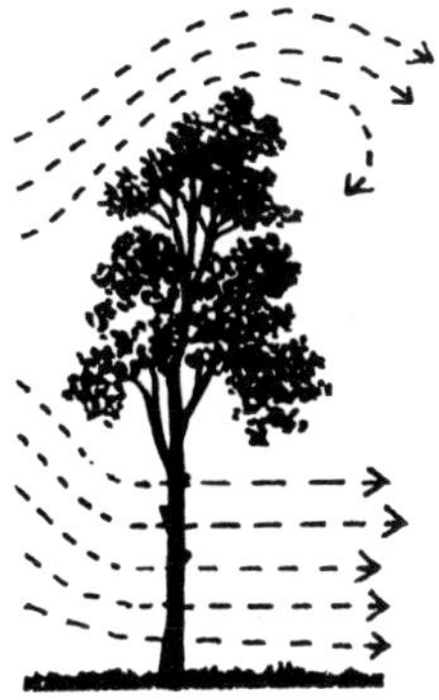

The spacing will also vary with the relative importance of the protective versus productive purposes of the windbreak. Where the products of windbreaks have a high priority, then land users may favor a greater number of shorter strips and a higher proportion of small trees and shrubs which provide products such as fodder and fuelwood. If the by-product is timber, the height of windbreaks and the intervals between them might be increased. Where the overriding interest is to protect valuable crops, farmers may try to keep the windbreaks as tall and as far apart as possible to obtain the most protection for the least amount of cropland devoted to trees.

The woody understorey and herb layer should be well established and maintained, both to prevent uncontrolled traffic through the trees and to maintain soil cover against the wind. One common design mistake can actually make matters worse. If the lower level is left open while the upper level is too dense, the result can be serious damage to crops close to the downwind side.

Roads and paths require careful planning and should not be permitted to cut straight through a windbreak. Where a path is essential,

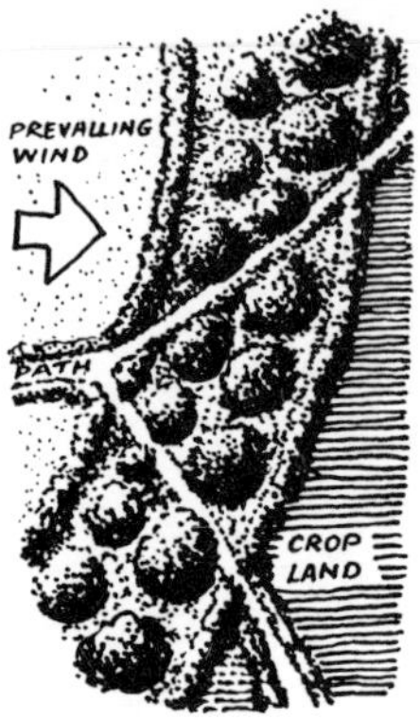

it should cross the windbreak at an angle. Place the path where it is convenient for local people and herders or they will make their own.

Where lateral root growth of trees may interfere with field crops, a deep trench may be dug along the tree belt to cut the roots. At the Central Arid Zone Research Institute in India, a trench 40 centimetres wide and 60 centimetres deep is dug about 1 meter away from the tree line. It is important to make sure that the trench will not discharge water which might cause erosion. Young tree roots near the surface can also be cut by deep tilling the adjacent cropland.

SPECIES

The most effective windbreaks provide a semi-permeable barrier to wind over their full height, from the ground to the crowns of the tallest trees. In general, trees with narrow, vertical growth are ideal for windbreaks to minimize the land removed from crop production. Since the shape of the windbreak changes as trees grow, it is usually necessary to mix several species with different growth rates, shapes and sizes in three or more rows. Some fast-growing species should be used to establish the desired effect as rapidly as possible. *Eucalyptus, Cassia, Prosopis, Leucaena* and *Casuarina* species are often planted for this reason. However, no tree will grow rapidly if it is not well adapted to the environmental conditions of the site. In addition, many fast-growing species do not live as long as slower-growing trees. Fast- and slow-growing species should thus be mixed to extend the useful life of the windbreak. Mixing species also provides protection against diseases or insects that can easily infest and destroy single-species stands.

AZADIRACHTA INDICA

Diversifying the species in windbreaks can also provide a wider variety of useful products to local users. A fully developed windbreak can yield wood, fruit, fodder, fibre and honey for sale or home use. Where animals are allowed to graze nearby, at least some of the lower, outer trees or shrubs should be unpalatable, while fodder species may be planted in the centre or along an inside edge where they are not exposed to animals but can be harvested and fed to livestock kept elsewhere. *Azadirachta indica* (neem) has been planted in windbreaks in Niger because its unpalatable leaves protect it from damage by livestock. In other areas, however, goats browse on neem leaves, possibly due to different chemical properties associated with different varieties of this species.

Some trees, such as *Azadirachta indica* and species of *Casuarina* and *Eucalyptus,* should be used with caution in windbreaks. *Eucalyptus* should not be used alone, as it has a sparse understorey and may have negative effects on water availability and crop productivity nearby. Trees with dense or spreading canopies, such as *Azadirachta indica*, should be avoided near cropland unless they can be trimmed or harvested frequently. The species selected must fit together as a group into a larger overall design that fits into the local landscape and land-use system. In some cases people have planted successful windbreaks using such unlikely trees as *Anacardium occidentale* (cashew) and indigenous *Acacia* species.

While diversity is important, several considerations limit the choice of species. Trees and shrubs must meet the form, size and growth-rate requirements for the windbreak as well as the production priorities of the local people. Environmental problems such as insect pests (especially termites), wild and domestic animals, poor soil and drought also narrow the choice of species.

Water management is important, especially during the establishment phase. In the Turkana District of northern Kenya, trees were established successfully with only 180 mm of annual rainfall and without watering by using microcatchments (see section 5.4). If microcatchments cannot be constructed, hand watering or irrigation should be planned. The importance of watering and species choice was illustrated at a site in Libya with only 100 to 200 mm annual rainfall. Seven different tree species were watered once when planted and again 2 weeks later. Afterwards, they were watered 2 times a month, 4 times a month or not at all.

The percentages of young trees of three species surviving after 8 months are as follows (United States Agency for International Development (USAID), 1987):

Species	No Water	Water Twice/Month	Water 4 Times/Month
Acacia cyanophylla	84	4	100
Casuarina equisetifolia	30	90	90
Eucalyptus camaldulensis	36	60	90

In the Near East, seedlings are usually watered 5 to 6 times a year for 2 to 4 years after planting. Experience from Kenya's Turkana region shows that enough water must be provided to soak deeply into the soil. Otherwise, the trees develop roots near the soil surface instead of growing towards the water table, making them ill-equiped to survive when watering is stopped. See the discussion on watering in section 5.4.

ESTABLISHMENT

While windbreaks can be established by direct seeding, it is best to use cuttings or seedlings whenever possible, at least for the upper-storey trees. Where browsing damage by wild or domestic animals is a problem, farmers may establish an outer living fence of densely spaced cuttings of unpalatable or thorny species. This requires considerable planning since the fence should be established well before the windbreak to provide adequate protection. Temporary thorn or woven fences may also be used. Where small windbreaks follow property lines, it may be possible to incorporate existing living fences as the first row on one side. The importance of protecting seedlings from browsing during establishment may also influence decisions about spacing, since it is easier to protect a few windbreaks of tall trees rather than several windbreaks of shorter trees.

Physical protection may not be the most effective method. Where livestock are herded, the herders may be instructed to keep their animals away from the seedlings. In Niger, young trees were successfully protected by employing watchmen who also acted as local extension agents and explained to the animal owners, who were almost all local farmers, the benefits of protecting the young trees.

Trees may also be damaged by harvesting their products prematurely. The Forestry Research Institute of Nigeria reported that the *Azadirachta indica* trees used in their shelterbelt trials were seriously damaged by the inhabitants of a nearby village. Due to their high local value as a medicine, the branches of young seedlings were cut repeatedly. Community education and participation in decision making are clearly indispensable if a windbreak project is to overcome local environmental and social constraints. The local people may know better than outside development workers the most effective means to protect young plants.

MANAGEMENT

A windbreak must be managed to maintain correct density and structure, with harvesting carried out with care. In Niger, farmers pollarded windbreaks to maintain the proper canopy density and to provide wood products. They harvested wood from a double row of 7- to 9-year old *Azadirachta indica* trees, spaced at 5 x 5 metres, either by partially pollarding every tree (trimming branches overhanging the fields), by fully pollarding every fourth tree at 2.5 metres above the ground or by fully pollarding every tree in one row. In all cases, wind reduction of 20 to 30% was maintained.

ANTICIPATED BENEFITS

Although very little information is available on the quantity of wood that can be harvested from trees growing in windbreaks, some preliminary results have been encouraging. In trials conducted in the Majjia Valley of Niger, with favourable soils, a relatively high water table and 425 mm mean annual rainfall, *Azadirachta indica* trees under proper management yielded between 3 and 7 kg of usable fuelwood a year averaged over the lifetime of the tree.

Based on this calculation, a double-row windbreak 100 metres long, with trees spaced 4 metres apart within rows, would provide about 250 kg of wood a year. A windbreak of this length would protect about one hectare of cropland. If a family of five protected six hectares of cropland with windbreaks of this design, they would be able to meet their fuelwood needs for the year. Pollarding these same windbreaks every 4 years would provide construction poles and wood valued at US$800 per kilometre of windbreak. Remember, however, that wood cannot be harvested for several years after the windbreak is planted.

Anacardium occidentale (cashew) trees, used in a windbreak in Senegal, yield nuts which, although not sufficient in quality and quantity for large-scale commercial production, provide an important addition to local diets. *Acacia scorpioides*, planted in windbreaks in Niger, produce seed pods used for traditional leather tanning. Since there is a steady market for this product, the windbreaks make a modest, but much appreciated, contribution to the local economy. In other cases where windbreaks have been established with *Prosopis*, seed pods are collected daily for supplemental livestock feed and some are sold on the local market.

The reported effects of windbreaks on crop yields vary considerably. In some cases grain yields increased significantly; in other cases the competition for water and light, the land 'lost' to tree planting or changes in the microclimate were slightly detrimental. The effect on yield clearly depends in large part on the design of the windbreak and on the particular crop and environment involved. Because of this, the tree products obtained from windbreaks and the long-term benefits in

BUTYROSPERMUM PARKII

MARKHAMIA LUTEA

terms of soil conservation should be emphasized when discussing the benefits of windbreaks.

In northwest China, multi-row shelterbelts of *Paulownia* have been planted to stop the spread of deserts. A decrease in wind speed of 21 to 55% was measured in the protected area, along with a 12.5% increase in air humidity and a 19.4% increase in moisture in the top 50 cm of the soil. The introduction of shelterbelts also moderated maximum and minimum temperatures, increased crop yields and resulted in significant wood production. In some cases, however, maximum high temperatures *increased* due to windbreaks, resulting in damage to crops. These results emphasize the importance of carefully considering design and management factors before planting windbreaks.

In the Sahel, windbreaks seem to have a positive effect on the crop yield of protected fields. During a year of above-average rainfall in the Majjia Valley of Niger, sorghum and millet yields in fields protected by *Azadirachta indica* windbreaks were as much as 23% higher than in unprotected fields nearby. With rainfall 46% below average, yields were still about 16% higher with windbreaks. In years of poor rainfall, even small differences in crop yields were significant for the local population. With mature windbreaks—over 10 years old—an estimated 17.4% of land was lost to crop production due to shading, but this area produced wood and fodder.

In the Gezira irrigation scheme in Sudan, planners predict that shelterbelts could increase the yield of existing irrigated fields and save water to irrigate additional land. The cotton yield is expected to increase by 6 to 16%, depending on the exact design and effectiveness of the shelterbelts, while a crop yield increase of 5% would be enough to cover all establishment and maintenance costs. Additional crop yield increases, plus wood, fodder and honey production and soil improvement, would be clear profit. Planners estimate that a family with 10 hectares would harvest 7.5 cubic metres of wood and 1 ton of fodder from shelterbelts each year, plus 0.41 ton of extra cotton. In addition, the savings in irrigation water would allow 2000 new families to join the scheme, each with an allocation of 10 hectares under irrigation.

COMBINATION WITH OTHER TECHNIQUES

Windbreaks can be combined with dispersed trees on cropland (section 4.1), such as *Acacia albida* on sandy 'millet soils' in the shrub savannah zones. A fairly tight (10 x 10 metre) grid of *A. albida* between windbreaks protects the soil against wind and evaporation. In areas with more rain and/or heavier soils, other species, such as *Butyrospermum paradoxum* (karite), *Markhamia platycalyx, Ficus* species or *Borassus* palms, could be used. See Table 2 in Appendix I for suggested windbreak and cropland trees.

Small windbreaks may be useful for local vegetable gardens and small irrigation schemes. In these cases, they could be combined with living fences (section 6.1). The design would have to take the small size of plots and their layouts into consideration: tree height would be less than in larger windbreaks designed to cover an entire plain or valley. Different species may also be appropriate: *Moringa oleifera*, for example, or *Gliricidia sepium, Markhamia platycalyx* or fruit trees. If irrigation water is available, the most productive and marketable fruit trees are likely to interest farmers. Passion fruit (*Passiflora edulis*) functions as a windbreak on small farms in Kenya, where it is trained along wire fences. Living fences may also be used on the outer edge of large windbreaks to help protect young trees from animals and pedestrians.

EXAMPLES FROM THE FIELD

The Niger Government, USAID and CARE are conducting a windbreak research and testing programme in the Maggia Valley, Tahoua Department. Since 1975, over 300 km of windbreaks have been established to protect over 3000 hectares of farmland. These originally consisted of double rows of *Azadirachta indica*, but this species was later replaced and/or mixed with *Acacia scorpioides* and *Prosopis* species. Researchers are compiling information on yields from individual trees cut under different management cycles, on crop yields with and without windbreaks and on the distribution of benefits.

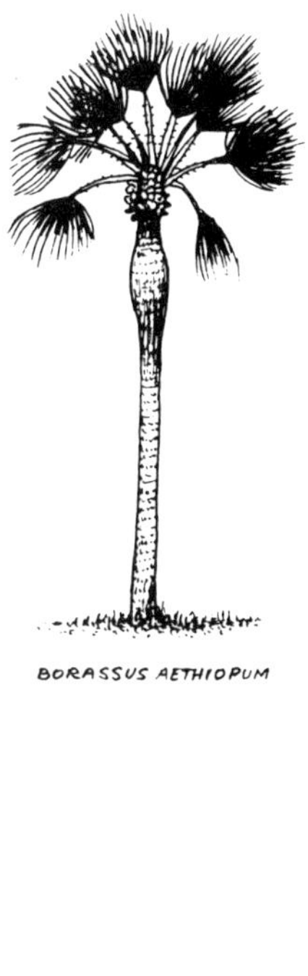
BORASSUS AETHIOPUM

Also in Niger near Maradi, indigenous species and low-input techniques were used in a windbreak demonstration. When land was cleared at a farmer training centre, strips of natural vegetation 10 metres wide were left every 90 metres facing the prevailing wind. These have since developed into diverse vegetation strips of local species. The entire training centre was surrounded by a strong and well maintained fence, illustrating the importance of protection from animals.

In irrigated areas along the southern coastal belt of Somalia, extensive windbreaks planted during colonial times still exist. These consist mainly of *Eucalyptus* and *Casuarina* species. Large-scale windbreak programmes are also in progress in central Senegal, in some cases using *Anacardium occidentale* (cashew).

In Tanzania's Singida District, farmers use *Euphorbia* hedges to protect croplands, while shelterbelts of *Grevillea robusta* are used to protect tea plantations in Kenya. However, insect pests that attack tea have been known to take refuge in these trees, reducing the popularity of shelterbelts. In Kenya's dry and subhumid savannahs, *Eucalyptus, Casuarina* and *Juniperus* species have been planted in windbreaks on large commercial farms and homesteads. While there is information on the performance of different tree species in these areas, little is known about the actual effects of windbreaks on crop yields and soil erosion.

Some farmers and researchers question whether shelterbelts might harbour tsetse flies. Experience in Kenya and Tanzania suggests that windbreaks need not shelter tsetse if they are well managed and the correct species are chosen. If the understorey is fairly open, the overstorey is high and the ground surface clean weeded, then tsetse infestation is not likely. However, the windbreak will be quite permeable.

SELECTED REFERENCES

Bognettean-Verlinden, E. 1980. *Study of impact of windbreaks in Majjia Valley, Niger.* Wageningen, The Netherlands: Agricultural University. 77 pp.

Darnhofer, T. (1983) Microclimatic effects and design considerations of shelterbelts. In D.A. Hoekstra, ed. *Agroforestry systems for small-scale farmers: proceedings of an ICRAF/BAT workshop.* Nairobi: ICRAF, pp. 95-111.

Dennison, S. (1986). *Majjia Valley windbreak evaluation: Niger briefing document.* Niamey: CARE, 7 pp.

FAO (1985). *Tree growing by rural people.* Forestry Paper No. 64. Rome: FAO, 130 pp.

Re-establishing trees around a watering place for cattle.

FAO (1986). *Brise-vent et rideaux abris avec référence particulière aux zones sèches.* Cashier FAO Conservation No. 15. Rome: FAO, 385 pp.

USAID (1987). *Windbreak and shelterbelt technology for increasing agricultural production.* Washington, DC: USAID, 219 pp.

Weber, F. and Hoskins, M. (1983). *Agroforestry in the Sahel.* Blacksburg, Virginia, USA: Virginia Polytechnic Institute, 102 pp.

WMO (1964). *Windbreaks and shelterbelts.* Technical Note No. 59. Geneva: WMO Secretariat, 188 pp.

6.4 Trees and Shrubs along Waterways and Floodplains

DESCRIPTION

Throughout dryland Africa, natural vegetation is often most dense along seasonal waterways, in floodplains or in seasonally flooded depressions. Even in arid areas, conditions near these water sources are generally favourable enough to support productive trees, shrubs and grasses that otherwise would not survive. Agroforestry and social forestry programmes can re-establish bands of woody vegetation and grasses on the banks of streams, lakes or seasonally flooded ponds or swamps where natural vegetation has been removed or degraded.

The basic objective is to protect fragile land along waterways and make it more productive. Vegetation can prevent the stream bank erosion that affects a great deal of productive land every year. The edges of swamps and seasonally flooded depressions (called 'mares' in West Africa and 'dambos' in southern Africa) can also produce a wide range of subsistence and commercial crops in gardens, woodlots and managed stands of indigenous trees and grasses.

Seasonally flooded land along streams and lakes is often dry for most of the year, but may have underground water close to the surface. Although unsuitable for many cropping systems, these sites can support productive, fast-growing trees, shrubs and grasses. A similar situation exists along irrigation and drainage canals. Woody plants and grasses can grow in these locations where little else would be productive or where annual cropping systems would destabilize the soil, causing erosion of the land surface and adding sediments to the water.

Agroforestry practices for these sites differ both in structure and purpose from techniques to protect open channels and gullies. In these cases, the water course, lake or pond is a given, with planting focussed on protecting and utilizing the margins. Physical structures are less important than in the case of gullies or water-control channels where changing the rate and path of water flow is usually a major objective.

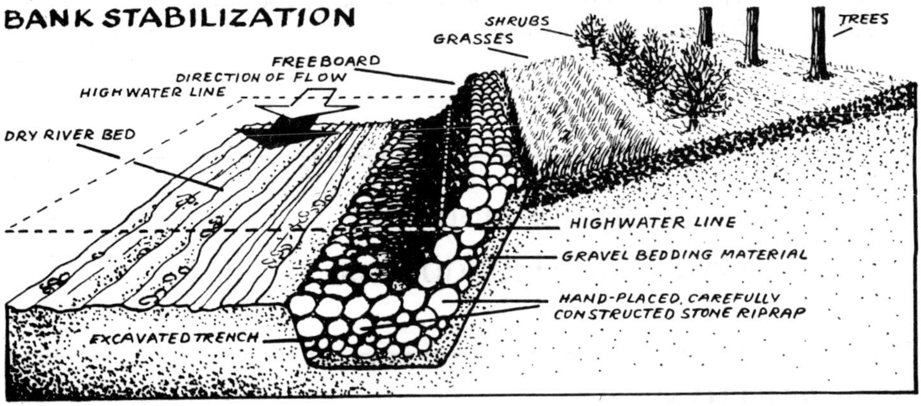

DESIGN

Several different agroforestry practices can be used on floodplains. These sites provide an abundant supply of water for plant growth and their soils are often fertile, with a high organic matter content. However, they also present special limitations that must be considered in designing agroforestry systems. In some situations, acid or saline soils may restrict the choice of species. Some plants may not tolerate the seasonal and annual changes in water level, both above and below the soil surface. Flooding may disturb plants in several ways, ranging from waterlogging to direct physical damage by running water, erosion of topsoil or deposition of course sand and gravel at the planting site.

Water may erode and deposit soil and rocks and cut new channels. People living near seasonally flooded areas know how risky it is to invest in permanent structures on riverbanks. Workers planning an agroforestry project should consult the local community and available technical information on the expected frequency and intensity of flooding, as well as droughts. Never forget that most water courses change over time.

Agroforestry practices must be compatible with local use of a water resource, both above and below ground. Floodplains are often considered to be public land, either by law or according to long-standing practice. Both people and livestock may use streams, lake margins and other water points on a regular basis. Local people usually value these sites highly and may have many customary access points to the water source. Even when dry, people may collect water form temporary wells dug in a riverbed or depression. Dry riverbeds are also frequently used as roads, footpaths or resting places, and it is important to stabilize the banks so that they can sustain such traffic. Plants established at these sites must be well protected by local agreement or they must be able to resist browsing, trampling and frequent cutting. In either case, it is a good idea to include well placed, well defined access routes to water when designing an agroforestry project on such a site.

Because of the proximity to water, these sites are often ideal for tree nurseries that require frequent watering. However, they may be difficult to protect from animals or thieves if they are in isolated locations.

SPECIES

Unlike many agroforestry practices, floodplain management often calls for separate, parallel strips of grasses, shrubs and trees, rather than a combination of all three in the same place. A strip of grass is especially important along the edge of fast-flowing streams.

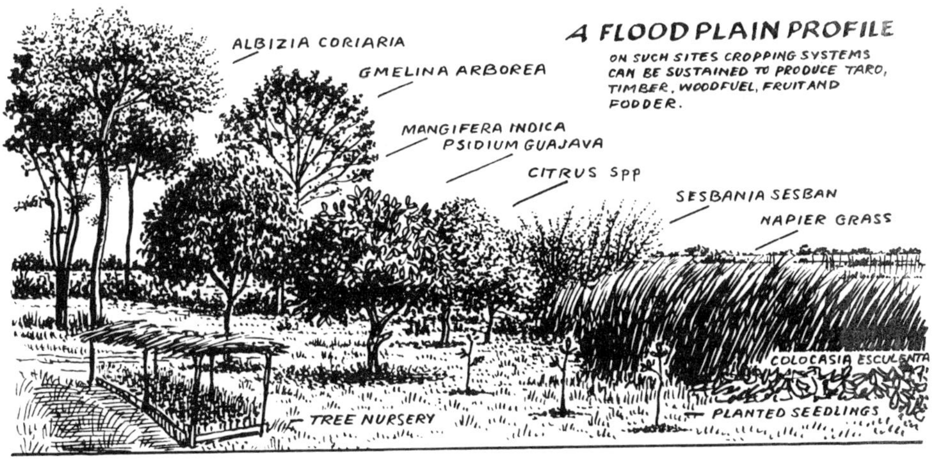

In India, trees have long been planted along canal bunds, usually 6 metres from the inner edge. Among the species most widely used are *Dalbergia sissoo, Eucalyptus* species, *Albizia lebbek, Azadirachta indica, Acacia nilotica, Acacia tortilis, Melia azedarach* and *Parkinsonia aculeata.* Establishment and management methods range from direct seeding with no watering and little weeding to planting out of tall seedlings in pits of 50 cubic centimetres and watering 4 or 5 times from the canal.

In some parts of the world, lakes and streams have been choked by the invasion of aggressive plants introduced onto the margins of standing or flowing water. Their seeds or sprouting pieces are rapidly dispersed by the water and colonize areas far from the original site. For this reason, it is best to take a conservative approach to the introduction of exotic species along watercourses and standing water. If possible, choose indigenous species known to survive in such sites without encroaching on either the water resource or neighbouring vegetation.

In many instances, fruit trees are a logical choice. *Phoenix dactylifera* (date palm), *Citrus limon* (lemon), *Citrus sinensis* (orange), *Mangifera indica* (mango), *Ficus* species (fig) and many other fruit or spice trees may be possibilities, depending on climate and soil conditions. *Casuarina* species have been widely used along irrigation and drainage canals in several countries, most notably in Egypt. While some *Casuarina* species have proven to be too aggressive, *Casuarina cunninghamiana* (river sheoak) fits well into the waterline environment.

If surface water from the stream or water body is a resource to be protected, then it is important not to introduce species that will invade and/or drain the area. Introduced plants must not deplete soil moisture or reduce stream flow, lake volume or groundwater reserves during the dry season. For example, *Eucalyptus* trees, which have been planted in some cases specifically to drain swamps, should be used with care, paying special attention to the variation in water utilization between species. On the other hand, where seasonal flooding occurs, plants must be able to tolerate waterlogging or even partial submergence. Those species known to tolerate periodic flooding and waterlogging are listed in Appendix I, Table 2. Trees that grow well along rivers are noted in Table 1 of the same Appendix.

MANAGEMENT

Trees are easily established along waterways or floodplains if the soil is moist and reasonably deep, which is usually the case unless the banks are steep or rocky. Planting by direct seeding or cuttings is usually preferable in stable sites, especially for fast-growing 'pioneer' plants such as *Sesbania* or *Gliricidia* species. It may be necessary to plant seedlings if woody plants and grasses must be well anchored before the next

season's flood, depending on the type of site and the specific position of the plants. This is a particularly important consideration along streams prone to flash floods, common in semi-arid areas. Much hard work can be washed away within a few hours.

Water causes erosion damage in a seasonally flooded forest.

In floodplains with wide fluctuations in water level, it is advisable to establish most new plants after the rainy season, when the soil is moist but not flooded. This is particularly true of those plants at or very near the water's edge. Higher up on the stream banks or lake shore, plants may be established at normal planting times, usually with the onset of the rains.

Plants established in multistorey gardens on waterways or floodplains may require intensive management. By contrast, hardy, well-adapted trees, shrubs and grasses planted in strips may require little or no maintenance. The most common and often most difficult management requirement in these sites will be protecting plants and their products from people and livestock.

ANTICIPATED BENEFITS

While trees may consume large quantities of water along waterways and floodplains, they may also provide shade and reduce wind across the waterway, resulting in lower water surface temperatures and reduced evaporation. Trees may also contribute directly to production or to environmental management and protection in floodplain sites. For example, *Ficus* trees have been planted around fish ponds: the fruit is eaten by the fish and these are then harvested by farmers. In Kenya, unusually low populations of aquatic snails—vectors of the parasitic disease, schistosomiasis or sleeping sickness—were observed where *Eucalyptus* species grew along waterways. Experiments indicated that water that had been in contact with the leaves of *Eucalyptus globulus, E. albens, E. robusta* or *E. microcorys* was fatal to snails. *E. saligna* and *E. grandis* had no such effect (Cheruiyot *et al.*, 1981). In Northern Kenya, *Balanites aegyptiaca* planted along rivers is also reported to eliminate snails.

Trees and shrubs planted in bands along streams and rivers or at waste-water outlets can also filter out undesirable substances from farming or agricultural processing operations, such as coffee factories, sugar mills or slaughter houses, before they reach the main channel. Where abundant moisture is available along water courses, most trees can be highly productive, tolerating minor changes in surface or sub-surface water levels better than agricultural crops.

COMBINATION WITH OTHER TECHNIQUES

Tree planting along waterways or floodplains can easily be combined with windbreaks, especially on irrigated sites (see section 6.3), with protecting and stabilizing channels (section 5.5) or with multistorey gardens (section 4.4). Lake shores, pond margins and seasonally flooded depressions are often stable enough to support multistorey gardens with an abundance of tree crops.

EXAMPLES FROM THE FIELD

A number of 'mares' in the Sahel have benefited from tree planting around their shorelines; sometimes this has been in connection with the development of small-scale vegetable gardens. Several irrigation schemes in Africa have incorporated trees along canal lines, for instance on the Senegal River, the Niger River, around Lake Chad, at sites in Sudan and in valley-bottom development schemes in Kenya and Tanzania. The Bura irrigation scheme in Kenya incorporates fuelwood trees under irrigation with 'waste water'. In Kenya's Siaya District, farmers have planted *Leucaena leucocephala* and *Sesbania sesban* in floodplain gardens and have established tree nurseries on the edges of swamps.

Planting along water lines has proven to be an important component of China's 'Four Sides' tree-planting program. This effort has proven beyond doubt that there are many places where trees can be planted in the rural landscape without competing for agricultural land. Tree planting has been credited with a major role in eliminating the disastrous floods of the Yellow River.

SELECTED REFERENCES

Cheruiyot, H.K., Broberg, G., Wamae, L.W. and Wachira, T.M. (1981). Effects of eucalyptus leaves on the survival of aquatic snails. *East Africa Agriculture and Forestry Journal.* 46 (4): 77-80.

FAO (1978). *China: forestry support for agriculture.* Forestry Paper No. 12. Rome: FAO, 103 pp.

Gupta, R. (1980). *Plants for environmental conservation.* Dehra Dunn, India: Bishen Singh Mahendra Pal Singh, 247 pp.

National Academy of Sciences (1984). *Casuarinas: nitrogen-fixing trees for adverse sites.* Innovations in Tropical Reforestation Series. Washington, D.C.: National Academy Press, 118 pp.

6.5 Trees and Shrubs along Roads and Paths

DESCRIPTION

Areas along roads and paths are often available for planting multipurpose trees and shrubs. Trees in these areas provide shade, reduce dust on adjacent land and, if properly managed, provide wood, fruit, gum, oilseeds, bee and animal fodder and other useful products. In densely settled or treeless areas, roadside plantings may be a significant source of tree products for local communities. In areas where access to land and trees is limited, poor and landless people may derive an important income by harvesting tree products from roadsides.

Roadway drainage can cause major erosion damage and siltation in rural watersheds, but much of this can be prevented by incorporating simple channel-control structures and vegetation into road design and construction. Trees, shrubs and grass cover can stabilize freshly disturbed ground, and vegetation along roadways can protect cut and filled slopes and places where natural drainage patterns have to be changed. Along existing roads, trees, shrubs and grass can reduce the force of runoff, redirect drainage and stabilize drainage channels.

Drainage water from roads and paths can also be put to good use. In the Turkana District of northern Kenya, storm water is trapped in ditches along roads and the overflow directed to tree seedlings. This is similar to the 'Liman' system used in Israel. By stabilizing side slopes, roadside trees, shrubs and grasses may also contribute to safety, particularly along mountain roads. Careful spacing is required, however, so that trees do not create new traffic hazards.

Roadside planting, as such, is not an agroforestry practice. However, as in many other places in the landscape, roadsides present important opportunities to use agroforestry. Roadsides make an ideal demonstration site for new agroforestry species, planting and management techniques and plant combinations for use in other sites. Moreover, roadside agroforestry plantings can actually improve both productivity and physical stability in wide networks of roadsides and trails that amount to a substantial land area in most rural landscapes.

In cases where roadsides border croplands, careful planning can improve the compatibility of roadside trees with adjacent crops. Where some people have no access to cropland, food crops can be grown on roadsides. In this case, agroforestry workers need to combine agroforestry practices for cropland with the special design criteria required for roadside plantings to ensure site stability as well as productivity. In Africa, roadsides are also often used as sylvopastoral systems, with timber, fuelwood and/or fodder trees planted over grasses. Roadside fodder can be harvested or used for controlled grazing. Success will depend on a realistic system of access and management.

DESIGN

Tree-shaded footpath and road in a rural area.

Ownership and access to land and plant products along roads and paths need to be clarified early in the design process. Ownership is not al-

ways clearly or visibly defined, but in many cases specific arrangements do exist. Often the land and/or trees along roads and highways belong to the government. If so, it is important to determine where government jurisdiction ends. Some agencies have regularly scheduled maintenance and harvesting programmes to take advantage of the production potential of roadside lands. Substantial amounts of fuelwood and poles can be harvested, given the fact that roadside tree lines often cover considerable distances.

In other areas, trees and tree products along roadways belong to local communities. With some technical assistance from the government forest service, the communities earn a substantial revenue from the sale of fuelwood and poles. In yet other places, individual trees belong to individuals or families who may have helped to plant them. Here, the distribution of benefits may be more complex.

USING ROAD DRAINAGE

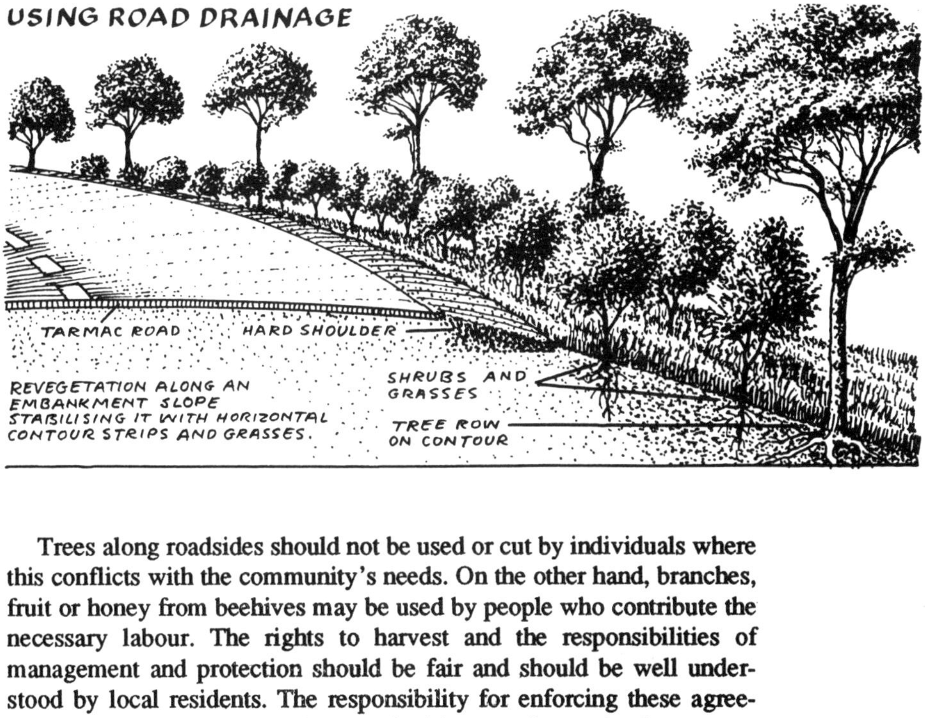

Trees along roadsides should not be used or cut by individuals where this conflicts with the community's needs. On the other hand, branches, fruit or honey from beehives may be used by people who contribute the necessary labour. The rights to harvest and the responsibilities of management and protection should be fair and should be well understood by local residents. The responsibility for enforcing these agreements usually resides with local authorities or adjacent land owners.

Most road designers focus on engineering and construction aspects and fail to consider the introduction and use of vegetation along the roadside. As a result, adjacent land users are exposed to problems of gullying, flooding and, on valley floors, sedimentation and destruction of productive farmland. Poorly planned drainage and lack of vegetative cover may also damage the road itself, resulting in blockage by mud slides, undermined sections or a rough, eroded surface. Once the damage is done, it is expensive to restore the road and neighbouring lands, but most of this damage can be prevented by proper attention to soil conservation at the design stage.

Several textbooks and field-construction manuals are available on the design of roadway drainage. Correct design reduces the erosion caused by concentrating runoff along roads. The sketches included here show how roadside trees and shrubs incorporated into physical control structures can prevent or reduce erosion and flood hazards. Apart from drainage and erosion control, roadside trees and shrubs are important visual and productive elements of the landscape. Decisions concerning species, placement, establishment, type of protection required, management and definition of user rights are influenced by traffic movement and the high accessibility of roadway sites.

The placement of trees along roads and footpaths must leave room for the safe passage of traffic, including people, animals and vehicles, and the trees must also be compatible with adjacent land use, which could include drainage ways, woodlands, cropland, pastures, home compounds, villages or markets. The spacing, size of mature trees and extent of the root system in roadside plantings should not interfere with the normal use of adjacent land. Roadside trees must also be able to tolerate the drainage and pollution conditions, competing vegetation and insect and animal pests particular to their site, as well as the use of adjacent lands.

When seedlings are planted along a trail or a road, it is difficult to envisage what they will look like fully grown. Fully developed trees on the inside of a curve can reduce visibility to such an extent that they hide oncoming vehicles. This potential danger should be taken into consideration when designing roadside plantings.

Placement is important even in flat, open terrain. Once trees have grown, they may not leave sufficient width for traffic to pass plus an adequate verge that drivers may use to avoid obstacles in the road, such as carts, people or animals. The distance of the first row of trees from the edge of the roadway must allow extra room for traffic to pass easily plus a safety margin.

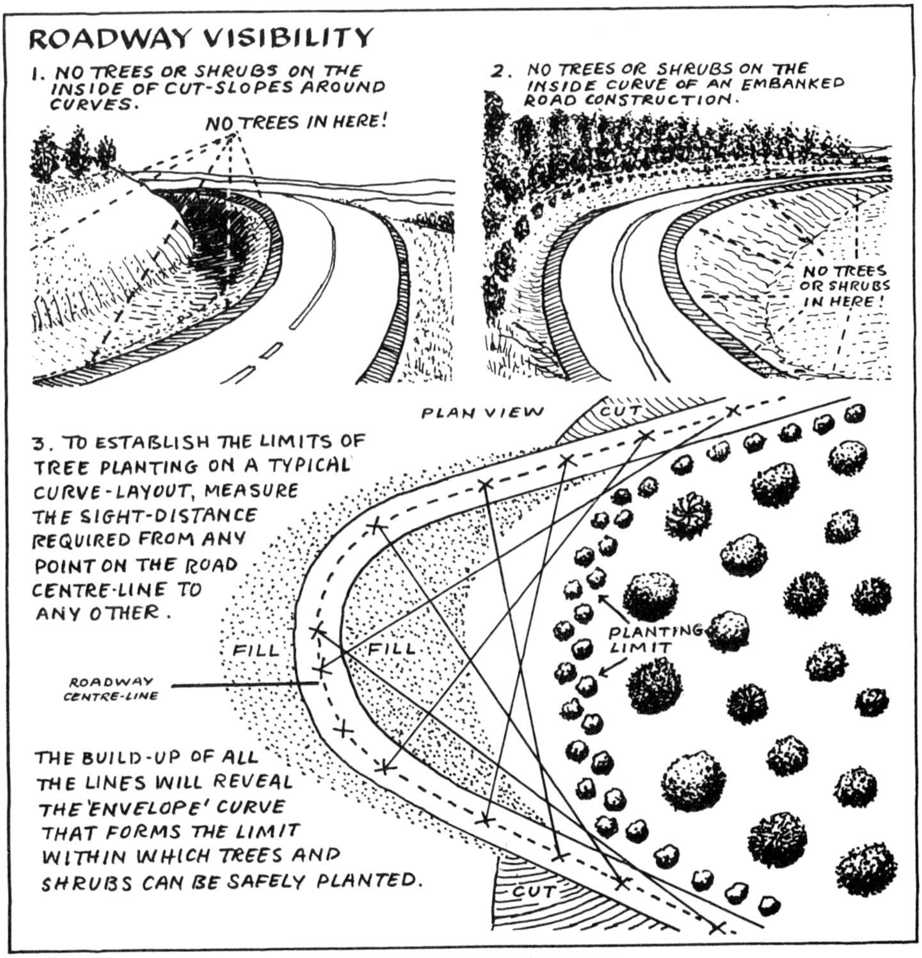

SPECIES

Many of the trees commonly used on urban and suburban roadsides have been planted for shade and decoration. Among the roadside ornamental species most widespread throughout dryland Africa are the *Jacaranda, Schinus molle* (pepper tree), *Delonix regia* (royal poinciana), *Melia azedarach* (Persian lilac), *Cassia siamea* and *C. spectabilis*. In semi-arid regions of northern Kenya, *Prosopis, Acacia, Cordia* and *Ziziphus* species are used. Table 1 in Appendix I shows some species appreciated as ornamentals. Most of these species can also be managed to produce fuelwood, fodder, small building poles, honey and, in the case of *Melia azedarach*, pesticidal extracts.

By the same token, many species known for their utilitarian value can be nested into the roadside landscape as part of a larger design for both shade and ornamental value. *Acacia* and *Prosopis* species and *Azadirachta indica* are especially good candidates for roadside sites because of their drought resistance and provision of fodder pods, gums, oil seeds and small poles.

If roadside trees and shrubs are to serve a variety of purposes, then the species must be chosen carefully to meet different requirements. If shade and erosion control are the main goals, then it may be best to choose trees that do not produce leaf fodder, fuelwood or other desirable products that local people would be tempted to harvest. If soil-erosion control plus fodder production are required, then the best species may be a pod-producing fodder tree with a broad crown and an extensive root system. If fuelwood is the major objective, followed by shade and soil stabilization, then separate species, or separate trees of the same species, may be used for each purpose. For example, some trees may be set aside for wood harvesting while others are left to develop fuller canopies and wider root networks in order to protect the soil. Table 2, Appendix I, suggests some species for planting along roads and paths.

MANAGEMENT

Protecting a roadside site by fencing the entire area is impractical and uneconomic. However, newly planted trees, shrubs and grasses are exposed to trampling, browsing and grazing by animals using the roads and trails. Given the public access and constant exposure to animal traffic, establishment of trees must be rapid and protection must be unusually effective to guarantee survival. Nonpalatable species may be used or physical barriers may be erected around individual trees or clumps of vegetation. Trees may also need to be pruned to remove dangerous overhanging branches or to prevent lower branches from blocking a pathway or impeding visibility.

ANTICIPATED BENEFITS

Shade is particularly important for draught animals carrying or pulling loads or for people walking, riding or bicycling along a road, especially in hot, dry climates. Where space is sufficient, two or three rows of trees and shrubs may be planted. Pedestrians and animals are likely to walk between the rows of trees, keeping the road free for faster-moving traffic. Cutting and harvesting operations must be arranged and timed so as not to jeopardize the provision of shade.

In many parts of Africa, raw materials from trees planted along roads are commonly used to prepare home remedies. Oilseeds, fruits, fodder and other products may also be harvested by adjacent landowners, government agencies or whole communities. Fodder production from these sites can be substantial. In the Kenya highlands, many women feed dairy cattle on fodder gathered from roadsides, with a supplement of Napier grass grown in small plots. These small dairy enterprises depend heavily on the productivity of natural vegetation along roads and trails. Well-planned combinations of fodder shrubs, trees and grasses in these highland environments could yield substantial amounts of high-quality fodder.

COMBINATION WITH OTHER TECHNIQUES

Depending upon the orientation of roads and trails, roadside plantations may also be designed to act as windbreaks (see section 6.3). In hilly areas where slopes exceed 10 to 15%, many of the principles ap-

Two rows of trees provide a shaded and protected right-of-way and yield fuelwood and poles.

plicable to ditches and terraces (sections 5.1 and 5.2) should be incorporated into plantings along roads and trails. This may be important for drainage and to stabilize entire slopes. Where roadside plantings border croplands or pasture, most of the considerations for boundary plantings will apply (section 6.2). If villages and market places border the roadside, the design and management of roadside plantings should follow the guidelines for trees in public places (section 6.6).

EXAMPLES FROM THE FIELD

In Rwanda, a nationwide programme to plant rows of trees along roads and trails has been in effect for several years and similar efforts have been undertaken in many other countries. While in most other cases one or two rows of trees are planted, the Rwanda programme has established strips up to 10 and 15 metres wide, providing much better protection to the soil. A newly proposed forestry law in Rwanda has several paragraphs dealing specifically with roadside plantations, covering design, spacing and management of trees, as well as the definition of beneficiaries.

In central Senegal, trees along roads are harvested on a continual basis, mainly by pollarding. In Morocco, *Eucalyptus* trees have been pollarded along highways for many years. In semi-arid regions of Bijapur District, India, the highway department harvests and auctions pods from *Tamarindus indica*, while the less valuable, more irregular yield of oil seed from *Azadirachta indica* trees is left to landless gatherers from nearby villages.

In order to use available cropland to the fullest extent possible, public works departments in other parts of the world have begun to make special arrangements with farmers to cultivate the slopes of embankments, and in some cases even slopes excavated during highway construction. Only crops that provide good ground cover and enrich the soil are permitted, such as beans. In addition, farmers must plant hedgerows of particular tree and shrub species such as *Leucaena leucocephala*, which they must keep trimmed to certain dimensions. Concern for road safety and visibility in part determine the management recommendations for roadside trees. At the same time, local people generally appreciate the opportunity to use public land and roadside slopes are better protected under these arrangements than if they are left bare.

SELECTED REFERENCES

FAO (1977). *Guidelines for watershed management.* Conservation Guide No. 1. Rome: FAO, 293 pp.

Gupta, R. (1980). *Plants for environmental conservation.* Dehra Dunn, India: Bishen Singh Mahendra Pal Singh, 247 pp.

Oresdain, L.E. and Barrow, E.G.C. (forthcoming). Soil conservation along Lodwar-Kapuna road project. In D.J. Thomas, ed. *Proceedings of the Third National Soil Conservation Workshop.* Nairobi: University of Nairobi, Agricultural Engineering Department.

6.6 Trees and Shrubs around Houses and in Public Places

DESCRIPTION

The rural landscape would be far less pleasant and productive if it were not for a wide variety of shrubs, trees, vines, grasses and other plants around houses and home compounds, around schools, in market places and in other public areas. While this may not be agroforestry in the strict sense, trees in these locations are just as important to rural people and their products are just as valuable as if they were growing on croplands or pastures.

Trees around houses, home compounds and in public places are a tradition throughout Africa. Community meetings are often held under

An Albizia gummifera *provides generous shade for a community meeting place.*

an especially large, shady tree that may have special significance for the local people. Trees planted in home compounds may eventually form part of home gardens, small orchards, tree nurseries, or livestock enclosures as land uses change. Their contribution to the well being of rural households is often underestimated by outsiders. However, both formal surveys and informal reports indicate that rural people are keenly interested in planting trees around their houses and compounds. In rural areas of dryland Africa, trees are consistently better managed and survive better in home compounds, where they can be protected, watered and harvested with ease, than anywhere else in the landscape.

Trees in household compounds and public places are an important area of common interest among rural communities and national governments and international agencies who are interested in increasing the number of trees in the landscape. Even though trees in these locations may not be directly related to conservation issues or efforts to increase the production of tree products, they provide a focal point around which people can start talking and sharing information. In discussions on tree planting in home compounds and public places, agroforestry workers can listen and learn, ask questions and get answers that may provide clues on how to approach a wide range of resource-management and conservation issues.

Some of the most popular indigenous trees in Africa appear in home compounds and public places, under a wide variety of environmental and social conditions. Among the most common trees in these sites are *Tamarindus indica* (tamarind), *Mangifera indica* (mango), several *Ficus* species (fig), *Dobera glabra, Terminalia* species, *Combretum* species and several wide-crowned acacias, especially *Acacia tortilis.*

Carefully chosen and nurtured trees add beauty to a home compound.

Shopkeeper tending Nerium oleander *in a market town.*

Many of the tree species now present in public and home-compound sites grow slowly. Often they were not planted, but have been carefully protected and maintained. In some cases, the presence of such trees determined the location of homes, markets or places of worship in the first place. More recently, as deforestation has proceeded, purposeful planting in such sites has become a common practice. *Cassia siamea, C. spectabilis, Jacaranda acutifolia, Delonix regia* (flamboyant), *Melia azedarach* (Persian lilac), *Psidium guajava* (guava), *Citrus* species and *Caricapapaya* (papaya) are particularly popular. People are often willing to plant traditional, slow-growing species, knowing that their descendents will benefit from the mature trees.

An effort to introduce more trees into these settings will not result directly in more hectares reforested or farm soils improved. However, most people are likely to benefit far more over the short term from a few trees planted in their home compound than from centrally designed and executed tree-planting projects. Trees introduced onto the home compound may also serve to familiarize people with new species, with new uses and management practices for well-known species and with new plant combinations. The knowledge and the landscape-planning experience gained from public and home-compound sites may be applied to larger-scale agroforestry systems in cropland, pastures and other places in the landscape.

The relative importance of tree planting on public sites or on home compounds varies widely depending on land-use and settlement patterns, ethnic groups, age groups and gender. While public demonstration sites have not been successful in some areas, in other places the village square may have the best-tended trees in the area. Trees can be established and maintained in public places by well-defined, well-organized groups or by highly motivated individuals such as business people at the market place or permanent staff of local clinics.

DESIGN

Woody plants are selected and sited in home compounds and public places primarily to add to the comfort, beauty and utility of the site. The most specific criteria for species planted in these sites are negative: they should not produce noxious odors or irritating pollen; they should not attract or harbour insects or animal pests that might be harmful to people, domestic animals, stored food or household water supplies; they should not interfere with structures on the site, either through aggressive rooting systems, by dropping large, self-pruned branches or through growth forms that result in an unstable or unwieldy size and shape.

The positive criteria for species and planting arrangements in home compounds are more a matter of local taste and suitability for specific sites. Local preferences for tree species, use and placement in family living spaces or public meeting places vary from one group to another and sometimes even from one village to another. Trees in home compounds often repeat popular arrangements or species introduced into market squares, schools or government compounds. While the same species, spacing and management may not always be appropriate for both sites, there is enough overlap between the two to justify using public sites to display species and techniques for home compounds.

When a village tree-nursery project is started, people will decide for themselves which species they want to plant. While outside specialists might choose single-purpose ornamentals for home compounds, rural people in hot, dry regions may think first about shade. In other places, people may prefer to plant fruit trees or a living fence that can also be pruned to yield fodder for goats (section 6.1). Many people choose certain species of shrubs or vines because they keep away snakes, rodents or insects. However, scent and visual appeal are also appreciated and may make a difference in choosing between otherwise similar multipurpose trees. Many plants which outsiders consider merely decorative may provide fuelwood, poles, fencing, fibre, medicines or raw material for crafts. *Jacaranda* is an example of an ornamental tree that also provides shade and fuelwood.

SPECIES

Designing tree plantings for home compounds and public places generally does not pose major problems. The major issue is the choice

With care and protection, a Terminalia catappa *becomes a striking feature in a school compound.*

of species. Table 2 in Appendix I lists agroforestry species which are suitable for these sites. Rarely, if ever, will local people need or seek advice as to where in their compound to plant a well-known shrub or tree. However, people will require information on the special properties of new and unfamiliar plants. For example, some trees can damage foundations and interfere with wells. Other trees may give off irritating pollen (*Croton* species) or may grow too tall and unwieldy to be managed near houses (*Eucalyptus* species). Trees and shrubs may also host birds or insects which are unwelcome near the home (*Rauwolfia caffra*). The special characteristics noted in Table 1, Column 27, of Appendix I may be important in identifying suitable trees.

Given a choice, people tend to use species they have seen nearby. In areas where expatriates have imported exotic species, local farmers are now growing *Bouganvillea* and *Delonix elata* (flame trees). Considered exotic only a few years ago, many species of *Cassia, Delonix* and *Jacaranda* have attained widespread popularity. In areas that may be too dry for these species, people are willing to water them regularly, even if water has to be carried long distances. Also, household waste water may be poured out at the base of the tree.

Traditional values and preferences may also influence the choice of tree species for planting in home compounds. In agropastoral areas of Kenya, *Erythrina abyssinica* is often protected near home compounds for religious reasons. The same is true of many *Ficus* species. In North Yemen, the most popular tree in the mountain areas is an introduced pyramidal cypress that local people consider particularly beautiful.

Trees and shrubs in public places are rarely established by direct seeding; they are far more likely to be planted as large cuttings, large seedlings or even transplanted as shrubs or small trees. Most trees in these sites can grow to full stature with a minimum of pruning and shaping, though some of the smaller trees, shrubs and vines require shaping and training. Unlike trees established in cropland or pasture, woody plants in public places and home compounds are often intensively managed. They may be hand watered during dry periods or throughout the year, weeded, trained and pruned as required by each individual plant.

Trees planted in public places do surprisingly well in spite of heavy traffic. Market places, clinics and school yards are prime examples. In these sites, people use short pieces of barbed wire, metal strapping from shipping crates, rocks, bricks or woven mats to protect young individual trees. In some markets, the individual stall lessees are responsible for the trees next to them. In school yards, individual pupils may take over similar responsibilities, perhaps in connection with school gardens. Trees also receive special protection and care in many government office compounds.

Protection for young plants in public places is often limited to 'social fencing'—the general recognition that the plant should not be damaged—or to small individual structures of thorn branches, woven fibre, wood or wire. Sometimes a food crop is planted just around a seedling to indicate that the plant should not be disturbed. These methods are usually adequate to deter browsing animals, pedestrians, herds, carts and other traffic.

ANTICIPATED BENEFITS

While shade is often of prime importance, many other benefits may accrue from trees in public places. In the town of Lodwar in northern Kenya, people collect *Prosopis* pods to feed sheep and goats or to sell in the local market—if the children have not eaten them first. In many areas, fruit, pods or leaves are harvested from trees growing around houses. Some *Ficus* (fig) species planted in home compounds serve four different purposes: planted as stakes, they form a fence around the compound; their leaves are fed to livestock; their fruit provides a welcome addition to the family diet; and the young shoots or flowers are used to prepare medicines.

COMBINATION WITH OTHER TECHNIQUES

Tree planting in home compounds and public places does not lend itself to combination with other agroforestry techniques, except living fencing around and within home compounds. The other closest practice is planting trees, shrubs and grass along trails and roadways, described in section 6.5. Home gardens (section 4.4) can be seen as an intensification of tree planting in home compounds.

EXAMPLES FROM THE FIELD

Wherever tree seeds, seedlings or cuttings are available, people take some and try planting them at home. Where nurseries have expanded to include tree species suitable for home compounds, demand has often outstripped supply. In many towns, self-help groups and individual entrepreneurs have begun to produce seedlings of ornamental trees for sale to their neighbours.

Throughout Africa, home compounds are used as observation grounds for new tree species and as the first place, other than cropland, for active landscape planning. In hot dry areas, especially where thatch roofing has been replaced by metal roofs, tree species are usually chosen and placed to provide shade. For example, *Jacaranda* and *Cassia* trees are widely used for shade in Kenya's Machakos District.

From the dry lands along the Senegal River all the way to refugee camps in Somalia, trees grown in compounds are surviving remarkably well. They are in better condition than most trees in reforestation sites. In some towns in Mauritania, it is so dry that only *Parkinsonia aculeata, Euphorbia* and *Prosopis* species survive, planted around tents. In other cases, such as home sites in the towns of Turkana District, Kenya, only a few *Acacia* and *Prosopis* species grow. Yet trees planted in home compounds survive and grow well even in the driest sites.

Where conditions are more favorable, entire villages may be dotted with trees planted around homes and in public places. The ever-present *Azadirachta indica* in West Africa is one example. In eastern and

Mauritanian nomads in their seasonal camp, where only Parkinsonia aculeata *and* Prosopis *survive.*

southern Africa, *Jacaranda* and *Psidium guajava* trees are found in widely scattered locations: these are often the first tree species that people plant.

Many home gardens in the *miombo* woodlands of northern Zambia include *Psidium guajava, Citrus* species and *Manihot glaziovii* (ceara rubber), planted earlier as individual shade, fruit and ornamental trees. As the sites have acquired more permanence and definition, farmers have interplanted castor beans, cassava, pumpkins, beans, bananas and additional fruit trees. In some cases the skills acquired in raising and maintaining a few fruit trees in home compounds have been applied to establishing small orchards and living fences.

In a CARE agroforestry project in Kenya's Siaya District, some women's groups raised seedlings of a popular ornamental tree, *Terminalia catappa*, and sold them for 100 times the price of species used for fuelwood and fencing. This experience with ornamental trees encouraged the groups to build nurseries and to learn plant-propagation techniques. They later used their knowledge and their nursery facilities for other agroforestry efforts.

An anecdote here illustrates how people value trees in home compounds. An official in one African country went to inspect the trees that had been planted 2 weeks earlier on national tree planting day but found no trace of them. Upon inquiring, he learned from people in a nearby village that their neighbours had dug the trees out and transplanted them to their home compounds.

SELECTED REFERENCES

Storrs, A.E.G. (1982). *More about trees: interesting facts and uses of some common Zambian trees.* Ndola, Zambia: Forest Department, 127 pp.

Teel, W. (1985). *A pocket directory of trees and seeds in Kenya.* Nairobi: KENGO, 151 pp.

von Maydell, H.J. (1983). *Arbres et arbustes du Sahel: leurs charactéristiques et leurs utilisations.* Eschborn, FR Germany: GTZ, 530 pp.

CHAPTER SEVEN

AGROFORESTRY IN PASTURES AND RANGELAND

DESCRIPTION

The production of woody plants combined with pasture or rangeland is often referred to as a sylvopastoral system. The trees and shrubs may be used primarily to produce fodder for livestock or they may be grown for timber, fuelwood, fruit, pollen and nectar for bee fodder, or to improve the soil. There is a clear need for sylvopastoral systems in dry savannah zones, particularly to help meet wood and fodder demands throughout the year and to maintain fodder reserves through dry periods. These systems can also help to maintain the stability and fertility of grazing lands and to reverse trends toward land degradation and desertification. The rehabilitation of highly degraded sites may require structures for soil and water conservation as well as planting, protection and management of vegetation (see sections 5.1 and 5.4).

Valued as fodder and browse, Acacia tortilis *also provides welcome shade in the dry season.*

The spacing of trees and shrubs can vary widely. For example, in subhumid areas coconut (*Cocos nucifera*) trees for commercial production are evenly spaced in pastures for dairy cattle. Trees, shrubs and grasses may also be grown on contour lines or in patterns predetermined by soil and water conservation structures in degraded pastures. By contrast, *Acacia* trees, which produce edible pods for livestock, are often widely dispersed in dryland pastures of local grasses.

Sylvopastoral production systems are not new to Africa. In fact, many pastoral and agropastoral peoples throughout the continent have traditionally used and managed woody plants in savannah grazing lands to produce fuelwood, fodder, building poles and other products for sale and domestic use. Browse trees and shrubs in African savannahs often have higher crude-protein and mineral content, and sometimes higher dry-matter digestibility, than the associated grasses, particularly during the dry seasons. Due to the highly irregular rainfall of semiarid areas and virtual disappearance of nutritious grasses during the dry seasons, trees and shrubs are an essential part of the pastoral environment. In the driest areas, there is often no grass whatsoever except for a short flush after the annual rains. For example, browse from woody plants was found to constitute 92% of dry-season goat fodder in Kenya's Turkana District.

Nomadic groups often choose their dry season grazing lands based on the abundance of trees and shrubs with green browse and high-protein pods. In fact, pastoralists usually show more interest in managing existing browse and establishing special fodder reserves than in cultivating crops. Pastoral people also often rely heavily on gathered foods from the savannah and forests along waterways. For example, *Balanites aegyptiaca, Tamarindus indica* and *Acacia tortilis* are all prized and protected in lands inhabited by the Pokot people in northern Kenya. In at least one instance, they have planted *Balanites* trees near a favourite water point in order to assure a ready supply of fruit for clan gatherings. In *Acacia* woodlands along the Runde River in Zimbabwe, agropastoralists and farmers selectively cut older *Acacia tortilis* trees for timber, leaving newly mature, pod-bearing trees and younger browsable trees to produce dry-season fodder. Agropastoralists and farmers in the savannah lands of eastern Kenya maintain *Terminalia brownii, Combretum* species and *Acacia tortilis* in pastures and grazing lands to provide leaf fodder and pods for their goats and draught oxen during the annual dry season, as well as during more prolonged periodic droughts.

The intensive management or purposeful planting of woody plants in dryland pastures is less common. This practice is widespread in the humid highlands, where land is scarce, but even in these areas it is less common than the separate production of wood and fodder in woodlots, boundary plantings, fodder lots or contour strips of tall grasses (see section 4.2). However, community-based researchers have documented a few promising initiatives in the rehabilitation of wooded savannahs. Farmers and livestock owners in Zvishavane District, Zimbabwe, have identified over 30 woody browse species for planting and/or protection in savannah grazing lands near their villages.

The main obstacle to the production of fodder trees in pastures and rangeland is the need to restrict animal access and grazing until the trees and shrubs are well established. On particularly productive sites, the short-term opportunity cost of excluding livestock can be high. If the site is degraded, little grazing may be lost, but few woody species may be available that can tolerate the degree of erosion, compaction and poor fertility characteristic of such sites. Unless there is an acute shortage of fodder, the limited return to be expected from degraded land provides little motivation to invest in protecting and rehabilitating a degraded site, particularly if a great deal of expensive fencing is required.

DESIGN

Agropastoral systems may occur in large expanses of savannah grazing land or in small, scattered pastures. In savannah areas, it is often feasible to improve fodder production as well as the condition of soil and water resources through selective protection and management of the most desirable of the existing trees and shrubs. Natural regeneration of these plants can be encouraged through rainwater harvesting with microcatchments (see section 5.4), as well as protection from grazing animals. In addition, it may be possible to ensure sustainable use of these lands by defining and enforcing clear, secure rights of use and access. For example, the Njemps people in Kenya's Baringo District asked their leaders to solve disputes concerning the use of trees and to allocate rights to harvest *Acacia tortilis* pods during the 1985 drought.

Pastoralists usually have a strong stake in maintaining fodder trees, and often have customary rules regarding their use. For example, the Turkana people of northern Kenya recognize individual rights to harvest specific trees and their products in woodlands along the Turkwell River and the Barabaig people of northern Tanzania adhere to well-defined rules concerning tree use and access by clans and villages. If tree ownership or use rights are well defined and legally supported, then herders and farmers can better protect the trees in their grazing lands against outsiders in search of charcoal, timber or cropland, as well as against their more short-sighted neighbours.

In farming areas where land for grazing is in short supply and existing pastures are scattered and degraded, farmers still have a number of options. They can interplant herbaceous fodder in cropland or rotate fodder-producing fallows with food crops. They can also increase fodder production dramatically by converting existing pastures to a multistorey system of fodder trees over grass and herbaceous legumes. If timber or fuelwood brings a good price, they may choose to plant multipurpose trees or to plant a mixture of timber and fodder trees into the overstorey.

Production systems based on the use of trees and shrubs in pastures may be established or improved by protecting and managing existing woody plants and encouraging their natural regeneration; by selectively removing less useful species, such as *Acacia reficiens* and *A. nubica*, and protecting and managing the valuable species; by planting new trees and shrubs in pastures; or by selectively cutting less desirable trees and shrubs in woodlands or forests and planting grasses and herbaceous legumes in the clearings beneath the remaining woody plants.

The pattern and spacing of tree establishment is determined in part by protection requirements. For example, clumps of woody plants in pastures are easier to protect than the same number of trees and shrubs planted in lines or dispersed throughout the pasture. This is especially important for young trees if there is not enough land to exclude animals from the pasture over a long establishment period. On sloping or degraded sites, both tree planting and the introduction of conservation structures should follow the contours to conserve soil and water. Evenly dispersed trees give the best results when larger trees are planted for pods or when leafy fodder trees are also intended for nitrogen fixing and site improvement. In the latter case, spacing usually ranges from 5 x 5 to 10 x 10 metres or more for large trees in dry areas and 2 x 2 to 3 x 3 metres for shrubs and small trees.

Water- and soil-conservation structures may be required to establish new trees, shrubs and grasses and to improve the growth of existing vegetation. The structures most often used in dryland pastures are

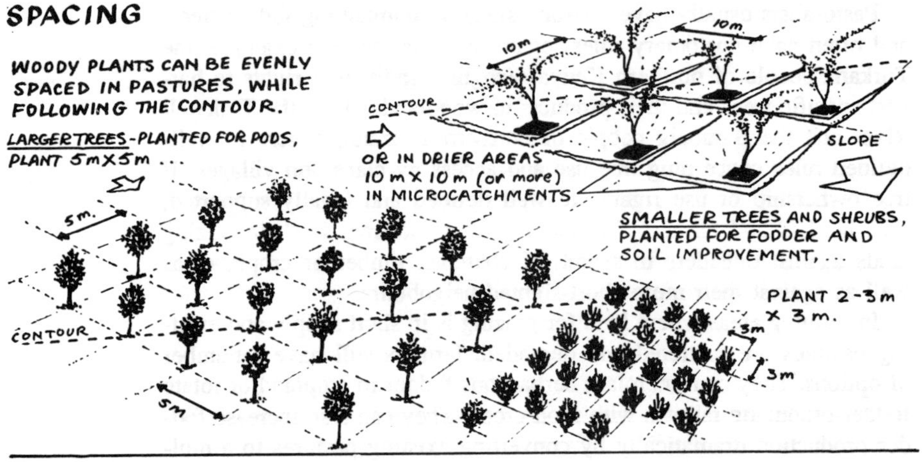

v-shaped or semicircular (crescent-shaped) microcatchments, ditches, ridges and pits. These structures are often temporary and are allowed to fill in or flatten out as the vegetation takes over the role of soil and water conservation.

Small earthwork structures are quickly constructed to enhance the survival of new plantings.

The type, size and position of conservation structures depend on the slope, the soil depth and texture, the amount of vegetation on the site and the intensity and frequency of heavy storms. The choice also depends on the availability of labour and on the priorities of land users—whether to reclaim sites or to intensify fodder production. Conservation structures are discussed in section 5.1 on earthwork structures and section 5.4 on microcatchments.

SPECIES

The choice of woody species for agroforestry in pastures and rangelands depends on local priorities and environmental conditions as well as on the degree of grazing pressure, the type of livestock present and the control of their access to the site. *Acacia* and *Prosopis* species are well suited to fodder-pod production in semi-arid rangelands, whereas *Leucaena leucocephala, Gliricidia sepium* and *Combretum* species are better suited for producing leaf fodder and fuelwood in improved pasture sites. In addition to fodder production, some tree and shrub species encourage grass growth underneath while others do not.

If it is difficult to control the access of livestock to the site where trees and shrubs are to be established, then the majority of seedlings should be nonpalatable timber or fuelwood species, along with a limited number of fodder plants that can be protected individually.

Other considerations include the ability of desirable species to regenerate naturally, the value of their products and their potential effectiveness for soil and water conservation. Species which are invasive and may become weeds are often readily established; these are good for some degraded sites, but hazardous for others. Check Table 2 in Appendix I for special characteristics of tree and shrub species.

UNESCO's Integrated Project on Arid Lands (IPAL) in the semi-desert country of Marsabit District, northern Kenya, has carried out tree-planting trials in an area inhabited by nomadic pastoralists. Where an initial period of watering was possible, the most successful introduced species were *Prosopis chilensis, Parkinsonia aculeata, Melia azedarach, Azadirachta indica* and *Carica papaya.* Although slower to grow, *Acacia senegal, A. tortilis* and *Salvadora persica* were the most successful of the indigenous species. Where watering was not possible in neighbouring Turkana District, *Prosopis* and *Acacia* trees were successfully established with only 180 mm annual rainfall using microcatchments.

Shrubs are particularly useful in grazing land, where their bushy growth and deep, spreading root systems allow them to produce fodder and food even during long drought periods. *Cajanus cajan* (pigeon pea), probably native to northeastern Africa, fixes nitrogen and grows on infertile soil, producing 'peas' for human consumption, pod husks and leaves for animal fodder and sticks for firewood, over a period of several years. *Cordeauxia edulis* (ye-eb nut) is a drought-resistant shrub that produces a nutritious nut—a staple food for some Somali nomads. Native to the Horn of Africa, *C. edulis* is an endangered species: it has been severely overexploited particularly during drought and famines. *Atriplex nummularia* (old man saltbush) is another drought-resistant shrub which is a useful source of forage in dry areas. *A. nummularia* is one of the most palatable species of the *Atriplex* genus and can tolerate saline soils and water. It is being tested in southern and northern Africa for forage production, with high yields reported.

The selection of species to protect in existing woodlands depends primarily on what species are available and on the value of their products. For example, in an *Acacia/Combretum* woodland in the savannah zone, a farmer might decide to remove *Commiphora* species but to use the cuttings to make a living fence, either to surround the site or to make an enclosure for livestock. *Combretum molle, Terminalia brownii* and *Acacia tortilis* trees of various sizes could be maintained for wood and fodder production as well as for site improvement.

In other places, there may be opportunities for collecting resins or gums from naturally occurring *Commiphora* (myrrh), *Boswellia* (frankincense) or similar trees, as practiced in Ethiopia, Sudan and Somalia. In an area with a good market for incense or other resin products, *Commiphora, Boswellia* and other resin-producing trees may provide a valuable cash income and thus take priority over fodder trees.

Akamba farmers use a Commiphora *hedge to provide a secure cattle enclosure.*

See Table 2 in Appendix I for trees and shrubs which might be introduced or maintained selectively in pastures. The species listed in the Appendix could be useful additions to the plants already growing in your area.

MANAGEMENT

Management of sylvopastoral systems in drylands usually focusses on fencing, special protection of newly established plants—whether naturally regenerated or introduced—control of grazing and, in some cases, maintenance of microcatchments. Leaf fodder and edible pods may also be harvested on a seasonal basis. Small plots may be more intensively managed, with both tree and herbaceous fodder harvested and carried to animals.

Trees may be protected through social agreement, rather than physical measures such as fences. Traditionally, specific areas are often reserved for grazing at certain times, based on control of livestock by the herders. The effectiveness of this approach was illustrated when an administrative official in northern Kenya forbid the felling or browsing of young *Acacia tortilis* trees: over 1000 hectares of *A. tortilis* woodland has regrown, with new trees now over 2 metres tall.

In Kenya's Machakos and Siaya Districts, tree regeneration in pastures is encouraged by propping up the main stem of young trees with poles out of the reach of livestock, especially goats. For tree species which tend to grow horizontally, such as *Acacia tortilis*, this practice reduces the time required until the crucial growing tip is tall enough to be safe from browsing animals.

ANTICIPATED BENEFITS

One of the major advantages of combining woody plants with pasture is the production of two useful products in the same space. In cases where valuable timber or other commercial tree crops can be grown with little or no effect on pasture, the economic incentive alone can easily justify the extra investment in planting, protecting and managing the trees. Livestock owners tend to judge the benefits of trees in terms of the dry-season fodder—pods and leafy browse—provided by woody plants. They measure the value of this dry-season reserve not by weight of fodder produced, but by the condition of their herds at the end of a drought or a long dry season. Woody plants in sylvopastoral systems can also provide many essential products for household use, including fuelwood, fencing material, construction poles, food, spices, fibre, dyes and medicinal herbs. The value of these products varies with local conditions, but it is likely to be substantial where they are in short supply. In addition, carefully selected woody plants may improve pasture growth by fixing nitrogen, stabilizing the soil and cycling soil nutrients. Another advantage is the shade provided by trees on rangeland. In most dry savannahs in Africa, shade at midday is essential both to livestock and to their herders.

Sylvopastoral systems—both traditional and experimental—are generally well adapted to savannah conditions. Maintaining trees in these areas can ensure the continued viability of pastoral and agropastoral land-use systems and halt or reverse the process of resource degradation where it occurs.

EXAMPLES FROM THE FIELD

Throughout Africa, trees and shrubs have been maintained in pastures and grazing lands for generations. However, well-documented examples of this practice are rare and many of the best reports are from other continents. For example, in the humid highlands of Costa Rica, farmers plant nitrogen-fixing *Alnus acuminata* in pastures at densities of 200 trees per hectare. In addition to a commercial timber harvest after 15 to 20 years, these trees help improve the condition and productivity of the pastures. In Australia, commercial ranchers in subhumid areas have also increased total return from land by planting timber trees in pastures.

Commercial ranchers in several countries have conducted some notable experiments in dryland savannah and grassland environments in recent years. *Prosopis juliflora*, formerly considered a useless weed, has improved livestock production when carefully managed in widely spaced, dense rows on ranchland in Texas in the southwestern USA. The high-protein *Prosopis* pods provide livestock with an ideal complement to the grasses, allowing a higher stocking rate and reducing the need for purchased feed.

In Kenya and Zimbabwe, several commercial ranches have successfully combined the production of domestic livestock and game animals, based largely on the efficient use and management of the diverse woody browse species in savannah lands. A comparative study of grazing schemes in different smallholder farming areas of Zimbabwe also attributed the fact that livestock were more productive and recovered more rapidly after drought at some sites to the abundance of woody browse plants in those areas.

In the dry shrub savannah of Kenya's Baringo District, one rancher established several plots of *Prosopis juliflora* or *Prosopis chilensis* in combination with grass on a grid of microcatchments prepared by heavy machinery. (The seeds of *Prosopis juliflora* and *Prosopis chilensis* have been mixed together by several seed suppliers in Kenya, leading to some confusion between these two species at many sites.) These plots successfully produced high-quality fodder, but protection of the trees has proven to be a major obstacle. The trial plots were surrounded by solar-powered electric fences, but these were ineffective against rats, which became a serious pest problem.

A fuelwood research project in the same area combined similar water-harvesting structures, but built by hand, with *Parkinsonia aculeata* and *Prosopis chilensis*. These trees have thrived under intensive protection, management and occasional watering, although if grass is introduced on the site there might not be enough water for both grass and trees. Local people have suggested planting the indigenous *Acacia tortilis* instead of these exotic species.

In Machakos District, KEFRI established grazing-land rehabilitation plots using a combination of agroforestry techniques. They used thorn fencing with living fences planted just inside. This project also included *Prosopis* trees planted for fodder on the inside of the fence line, grasses, trees and shrubs seeded directly on ploughed contour lines, widely dispersed trees planted in microcatchments or in deep holes throughout the plot and trees and grasses planted within and along pits and closed ditches.

Pits 0.5 metres deep, 0.75 metres wide and 2 to 3 metres long were used to harvest water and were planted with grasses and woody plants tolerant of occasional water-logging as well as drought. The banks formed upslope of the pits provided well-tilled, unconsolidated soil for planting larger fodder, timber and fuelwood trees, including *Cassia siamea, Parkinsonia aculeata, Prosopis juliflora, Acacia holosericea, Azadirachta indica* and *Melia azedarach*. Local community self-help groups provided the labour for site preparation and planting, while the project provided seedlings. After observing dramatic site recovery and enrichment during the trials, several farmers asked the self-help groups to carry out similar work on their degraded grazing land.

As part of a land-rehabilitation project in a semi-arid area of East Pokot, Kenya, several species of fodder trees were planted in microcatchments. The plots were fenced with thorn branches and local grasses were grown in between the trees. *Prosopis juliflora* performed especially well under these protected conditions, but the site managers noted the potential value of *Salvadora persica, Dobera glabra* and *Acacia tortilis* for unprotected land grazed by goats. Further experiments are now in progress in nearby Turkana District with these species plus *Acacia holosericea, Balanites aegyptiaca, Cordia sinensis* and others.

SELECTED REFERENCES

Amed, F.A. and Ahmed, A.E. (1978). Plant cover, nutritive value and carrying capacity of a range pasture in the low-rainfall woodland savannah of Western Sudan. *East African Agriculture and Forestry Journal.* 43 (3): 238-45.

Barrow, E.G.C. (1984). *NORAD's involvement in the forestry sector of Turkana District: past work and future prospects.* Lodwar, Kenya: Forest Department, 23 pp.

Boudet, G.C. and Toutain, B. (1980). The integration of browse plants within pastoral and agropastoral systems in Africa. In H.N. Le Houerou, ed. *Browse in Africa: the current stage of knowledge.* Addis Ababa: ILCA, pp. 427-32.

Lamprey, H.F. (1983). UNESCO's Integrated Project on Arid Lands (IPAL): a silvopastoral approach. In P. Huxley, ed. *Plant research and agroforestry.* Nairobi: ICRAF, pp. 129-36.

Scoones, I. (1988). *Community management of indigenous woodland project: Zvishavane and Chivi Districts, Zimbabwe.* Harare: ENDA, 105 pp.

Vietmeyer, N. (1985). In praise of shrubs. *CERES (FAO Review on Agriculture and Development).* 18 (2): 28-32.

Wilson, K. (1988). *Community management of indigenous woodland project: report to ENDA Zimbabwe.* Harare: ENDA, 115 pp.

PART THREE

TOOLS FOR AGROFORESTRY RESEARCH AND EXTENSION

This last section consists of reference materials, guidelines and examples—all tools to be used as needed. While the text contains references to specific parts of these appendices, readers are encouraged to explore this section as a whole in order to make the best use of the resources it offers.

Appendix I contains two sets of tables on trees and shrubs. Tables 1 and 2 focus on over 150 woody species that are well known in Africa. They give summary information on plant uses and products and on plant suitability for various agroforestry practices discussed in the text. Tables 3 and 4 provide a format to record and summarize information on plants already known and used by the local community.

Appendix II gives the full name and common synonyms for all the species listed in Appendix I. Each entry includes alternative Latin names as well as common names in several languages. A blank space is left for field workers to fill in the name(s) used in the communities where they work.

Appendix III provides sample questions, answer sheets and detailed instructions for the interviews and surveys described in Chapter 2. The first part covers the beginning of the survey process—the summary of the field workers' prior knowledge and the first general field visits. The second part contains sample questions and suggestions for general group and household interviews. These cover a wide range of topics and involve the local community directly in sharing information, ideas and decisions about agroforestry practices and the land-use system as a whole. The final part offers an example of a specialized in-depth interview, with sample questions and directions for conducting interviews, possibly in the context of joint work sessions.

Appendix IV is a sample format for summarizing the information and decisions recorded during the entire survey and discussion process. It provides one example of how field workers might condense the results of interviews and field activities in order to share and check information with the community. Once completed, these sheets could also help to focus group decisions on which agroforestry species and practices to implement, and where.

Appendix V contains a glossary of technical terms used in this book. Appendices VI and VII include a list of individuals and institutions that may be contacted for more information about agroforestry plus a list of the acronyms of organizations mentioned in this book. Finally, Appendix VIII brings together all the references listed in various chapters of the text plus others of general interest.

APPENDIX I: USES, PRACTICES AND CONDITIONS FOR MULTIPURPOSE TREES AND SHRUBS

Appendix I includes four tables which accompany this book but are printed separately in a larger format. Tables 1 and 2 are intended to serve as a guide to help you choose tree and shrub species that are suitable for your particular environmental conditions and that will provide the products and services you want. The information in these tables is based on experience in different places in Africa—either our own experience or the experience of others that we have learned about through personal communications or published reports.

In contrast to actual experience, ideas about the possible usefulness of various agroforestry species in different environments must be left to you—the reader—to observe, test, evaluate and judge. For this reason, Tables 3 and 4 are empty. They may be used to guide you in collecting and documenting information on the tree and shrub species you observe in your local area.

Some warnings associated with Tables 1 and 2 merit close attention:

1. Information on multipurpose trees and shrubs is often incomplete, conflicting, unreliable or specific to one particular site. So use these tables as indicators of species to consider, not as infallible prescriptions for your particular area and situation. We have been careful to use the most reliable sources available and to crosscheck the information they contain, but often important environmental and social variables are not recorded and techniques of measurement are not fully explained.
2. Plants react differently to different environments. A fruit tree may not bear fruit or a timber tree may be crooked instead of straight, depending on local conditions. The quality of a tree's products are generally best if the tree is grown in its most suitable environment. So if you plant a tree at the edge of its possible range, it may survive but not perform as well as expected.
3. Use information about your particular situation to evaluate and correct the tables. It is best to add local knowledge from your area and your own observations to the table and to make adjustments to match your situation.
4. The way people use a tree or shrub depends on their particular needs and preferences, on how well the plant grows in their area and on the other plants they may have available. In one place, a certain tree may provide poles that are perfect for building the local houses, whereas in another place the same tree may serve an entirely different purpose. Thus, the lists here only show how some people use different trees and shrubs, not necessarily how you might use them.

5. The information in Tables 1 and 2 is not complete—it is only a small selection of local and exotic species with some of their possible uses and environmental requirements. These tables are meant to suggest a few species, uses and environmental conditions for you to consider. Very likely, there are many other tree and shrub species growing in your area which you will also wish to take into consideration.

Table 1: Uses of multipurpose trees and shrubs

In this table, trees and shrubs are listed alphabetically by their botanical names down the left margin. The recorded uses of these species are listed on the top. For each species, the following codes indicate the importance of the use: 1 = very important; 2 = of secondary importance; X = used, but importance not known; - = negative effect or unsuitable; Z = poisonous product; * = poisonous product only under certain circumstances (processing often required); ** = related species have poisonous products; 0 = no information available. Column 27 lists any special characteristics which should be noted.

Table 2: Appropriate practices and conditions for multipurpose trees and shrubs

This table gives information on agroforestry practices, soils, regions in Africa and climatic zones that are reported to be suitable for different tree and shrub species. Four climatic zones are listed as follows:

1. **Highland subhumid**: altitude over 1000 metres, annual rainfall averages 500 to 1200 mm, minimum temperature averages -3° to 18°C
2. **Subhumid wooded savannah**: annual rainfall averages 900 to 1200 mm
3. **Semi-arid shrub savannah**: annual rainfall averages 500 to 900 mm
4. **Semi-arid tree steppe**: annual rainfall averages 150 to 500 mm.

Within these zones, local conditions may vary considerably, either in general or in exceptional years. In addition, many trees and shrubs grow well in different areas of several zones.

Table 3: Local uses of trees and other plants of interest for agroforestry

This table provides a sample form to record and summarize field information about how people use trees and other plants. Two copies are provided. Uses are listed across the top and space is left at the side to list plant species. You may wish to use these table as they are or to modify them, for instance by adding different uses for plants. You may simply mark the boxes to note which uses apply to each plant or to rank the importance of particular uses for each plant or the importance of particular plants for each use using numbers, for instance 1 to 3 or 1 to 5. Under some uses, such as food and fodder, you may wish to use letter codes to denote specific information. For example, you might write C (cow), S (sheep) or G (goat) to record which animal species eats a particular plant. You may also wish to note which information is based on your own observations and which is derived from other sources.

Table 4: Basic information on local plants, their surroundings and management

This table provides a format for recording 16 types of information about plant species observed in the field. Two copies are provided. They may be used as they are or modified, for instance by selecting only a few of the information categories listed or adding others. The following information might be included under the 16 headings:

1. **The local name** or names of each plant; note the language if there is more than one local language.
2. **The Latin name** of each species, if this is known; otherwise, plant samples or photographs can be used for future identification.
3. **Size at maturity and type of plant**; categories include large trees (20 metres or higher), small trees (10 to 20 metres), bushes or shrubs (less than 10 metres), climbing vines and non-woody understorey vegetation.
4. **Soil conditions** required or favoured by the plant include pH (acid or alkaline), fertility, texture (sand to heavy clay), presence of stones, depth, drainage (excessive drainage to water logging), ground water (near surface, fluctuating, deep) and salinity.
5. **Microclimate and light** refer to the plant's requirements for full sun or shade, temperature and humidity; response to other factors, such as frost, dew or wind, might also be noted here.
6. **Association and compatibility with other plants** might refer to other plants that tend to be found together with the species listed or to any positive or negative effects this plant appears to have on others growing nearby.
7. **Propagation and establishment** refer to possible and/or required practices, such as direct seeding, seed dispersed by animals, stumping (planting pruned stem-and-root cuttings), transplanting wild seedlings or propagating nursery seedlings in beds (for bare-root planting) or in containers; it may be useful to note whether natural reproduction requires fire, flooding or animal digestion and/or dis-

persal of seeds; recommended treatment of seeds or cuttings should be noted, such as nicking, washing in boiling water or applying rooting powder; also note any special requirements during establishment, such as well-tilled soil, deep planting holes, fencing, pest control, hand watering, shade, weeding or pruning.

8. **Growth rate** is a relative measure, but can usually be estimated using a well-known local plant as a standard for comparison; trees and shrubs can be rated as fast-, average- or slow-growing relative to this standard.
9. **Root form and development** include type and depth of root system, for example a shallow, wide root network or a deep tap root
10. **Plant management** refers to tolerance and/or requirements of established plants for such practices as pruning, pollarding, coppicing, grafting or budding.
11. **Pest management** refers to whether the plant is susceptible or resistant to specific pests (such as termites), whether it serves as a home for pests that may damage other plants, or whether it might function as a pesticide or pest repellant.
12. **Toxicity** includes information on what part of the plant is poisonous, under what conditions, to what species (people or livestock), and the toxic effects.
13. **Land unit** refers to the location of the plant on specific physical landforms, such as valleys, mountain tops, hill slopes, river beds, ridges, small depressions, gullies or plains.
14. **Land use** refers to a combination of plant cover and the use of the site, for example a home compound, a garden, a water point, a road or path, cropland, fallowland, pasture, rangeland, woodland or forest.
15. **Agroforestry practices** refer to a wide range of combinations of woody plants with crops, livestock and pastures, including: dispersed trees in cropland; contour vegetation strips; alley cropping; home gardens; improved fallows; trees and shrubs on terraces and small earthwork structures; trees, shrubs and grasses along roads, paths, erosion channels, gullies and waterways; living fences; trees on borderlines and boundaries; windbreaks; trees and shrubs in home compounds and public spaces; dispersed trees in pastures.
16. **Other information** may include whether the plant is thorny, forms dense thickets or is aggressive and likely to become a weed; local people may also wish to note other characteristics about specific plants.

APPENDIX II: MULTIPURPOSE TREE AND SHRUB SPECIES LIST

This appendix contains a list of multipurpose tree and shrub species that you may encounter in the field. Some have been mentioned in this book. The list is organized as follows:

Botanical name Authority (Family)
synonymous botanical names, if any
Vernacular or common names, where available, in English (Eng), French (Fr), Arabic (Ar), Hausa (Hau), Kiswahili (Kisw) and other languages, plus a blank space for the reader to add other local names.

This list represents a range of indigenous and exotic trees and shrubs grown in subhumid and semi-arid West, East and Southern Africa. It is based on the most recent authoritative sources on nomenclature available in 1988, but cannot be considered final. We hope that readers will add more species based on their own experience in the field.

Acacia albida Del. (Leguminosae subfam. Mimosoideae)
syn. *Faidherbia albida, Acacia gyrocarpa, A. saccharata*
Apple-ring Acacia, Winter Thorn, Ana Tree (Eng); Kad, Cad, Faidherbia (Fr); Haraz (Ar); Gao (Hau); Kababu, Mgunga (Kisw); ________ ( )

Acacia caffra (Thunb.) Willd.
syn. *Acacia fallax, A. multijuga, Mimosa caffra*
Hook Thorn, Cat Thorn (Eng); ________ ( )

Acacia dealbata Link
syn. *Acacia decurrens* var. *dealbata*
Black Wattle, Silver Wattle (Eng); Acacia Blanc (Fr); ________ ( )

Acacia decurrens (Wendl.) Willd.
syn. *Acacia decurrens* var. *normalis*
Wattle, Green Wattle (Eng); ________ ( )

Acacia eliator Brenan; ________ ( )

Acacia holosericea A. Cunn. ex G. Don; ________ ( )

Acacia karroo Hayne
syn. *Acacia capensis, A. natalitia, A. wirtella*
Mimosa, Sweet Thorn, Cape Thorn (Eng); ________ ( )

Acacia mearnsii De Willd.
syn. *Acacia decurrens* var. *mollis, A. mollisima*
Black Wattle, Tan Wattle (Eng); Mwati (Kisw); ________ ( )

Acacia melanoxylon R. Br.
Australian Blackwood (Eng); Mwati (Kisw); ________ ( )

Acacia mellifera (Vahl) Benth.
syn. *Acacia senegal* ssp. *mellifera, A. detinens, Mimosa mellifera*
Black Thorn (Eng); Kitr (Ar); Kikwata (Kisw); ________ ( )

Acacia nilotica (L.) Willd. ex Del. sspp.
syn. *Acacia scorpoides* var. *adstringens, A. arabica, A. scolioides, A. adstringens, Mimosa nilotica, M. scorpoides, M. arabica*
Egyptian Thorn, Scented Thorn, Prickly Acacia (Eng); Gonakié (Fr); Sunt (Ar); Bagarma (Hau); Mgunga (Kisw); ________ ( )

Acacia polycantha Willd. ssp. *campylacantha* (Hochst. ex A. Rich) Brenan
syn. *Acacia caffra* var. *campylacantha*
Hook Thorn, White Thorn, Falcon's Claw Acacia (Eng); Mkengewa, Mgunga (Kisw); ________ ( )

Acacia saligna (Labill.) F. Wendl.
syn. *Acacia cyanophylla, A. glauca*
Blue-leafed Wattle, Orange Wattle, Port Jackson Willow (Eng); ________ ( )

Acacia senegal (L.) Willd.
syn. *Acacia verek, A. rupestris, A. trispinosa*
Gum Arabic, Three-thorned Acacia (Eng); Gommier blanc, Vérek (Fr); Hashab (Ar); Dakworo (Hau); Kikwata, Mgunga (Kisw); ________ ( )

Acacia seyal Del.
syn. *Acacia stenocarpa, A. hockii*
White Whistling Thorn, White Galled Acacia, Seyal, Shittim Wood (Eng); Gommier (Fr); Thal (Ar); Mgunga (Kisw); ________ ( )

Acacia tortilis (Forsk.) Hayne subsp. *raddiana* (Savi) Brenan
syn. *Acacia raddiana, A. tortilis* var. *pubescens*
Umbrella Thorn (Eng); Tamadchi (Hau); Mguhga (Kisw); ________ ( )

Acrocarpus fraxinifolius Arn. (Leguminosae subfam. Caesalpinioideae)
Red Cedar, Shingle Tree, Indian Ash (Eng); ________ ( )

Adansonia digitata L. (Bombacaceae)
syn. *Adansonia sphaerocarpa*
Baobab, Monkey Bread Tree, Indian Cork Tree (Eng); Baobab, Calebassier du Sénégal, Arbre de mille ans (Fr); Tebeldi (Ar); Kuka (Hausa); Mbuyu (Kisw); ________ ( )

Adenanthera pavonina Linn. (Leguminosae); ________ ( )

Afzelia africana Pers. (Leguminosae subfam. Caesalpinioideae)
African Mahogany (Eng); ________ ( )

Afzelia quanzensis Welw.
syn. *Afzelia cuanzensis*
Pod Mahogany, Lucky Bean (Eng); Mkumbakusi, Mkonge, Mbarika (Kisw); ________ ( )

Agave sisalana Perrine (Agavaceae)
Sisal (Eng); Mkatani, Mkonge lume (Kisw); ________ ( )

Albizia adianthifolia (Schumach.) W.F. Wight (Leguminosae subfam. Mimosideae)
syn. *Albizia fastigiata, Acacia gummifera*
Flat Crown (Eng); Mchani-tsue, Mchani-mbao (Kisw); ________ ( )

Albizia lebbeck (L.) Benth.
syn. *Acacia lebbeck, Mimosa lebbeck, M. sirissa*
Mimosa, Siris, Women's Tongue Tree, Indian Walnut, Lebbeck (Eng); Bois noir, Langue de femme (Fr); Mkingu (Kisw); ________ ( )

Albizia schimperiana Oliv./Allo. Pet.
syn. *Albizia maranguensis, A. amaniensis*
Mimosa (Eng); Mduka, Mshai, Mkenge (Kisw); ________ ( )

Alnus acuminata Kunth (Betulaceae); ________ ( )

Alnus nepalensis D. Don; ________ ( )

Anacardium occidentale L. (Anacardiaceae)
Cashew Nut (Eng); Anacardier, Pomme Cajou, Pomme d'Acajou (Fr); Mkanju, Mbibo, Mkorosho (Kisw); ________ ( )

Annona muricata L. (Annonaceae)
Corossolier, Cachimanitier (Fr); Mstafeli (Kisw); ________ ( )

Annona senegalensis Pers. subsp. *senegalensis*
syn. *Annona chrysophylla, A. senegalensis* var. *latifolia*
Wild Custard Apple, Wild Soursop (Eng); Pomme cannelle du Sénégal (Fr); Gouanda (Hau); Mtomoko-mwitu, Mtopetope (Kisw); ________ ( )

Anogeissus leiocarpa (DC.) Guillemin & Perrottet (Combretaceae)
syn. *Anogeissus schimperi, A. leiocarpus* var. *schimperi, Conocarpus leiocarpus*
________ ( )

Arundinaria alpina K.Schum. (Graminae)
Mountain Bamboo (Eng); Mwanzi (Kisw); ________ ( )

Atriplex halimus L. (Chenopodiaceae)
Cape Saltbush (Eng); ________ ( )

Atriplex nummularia Lindley
Old Man Saltbush (Eng); ________ ( )

Azadirachta indica Adr. Juss. (Meliaceae)
syn. *Anteleae azadirachta, Melia azadirachta, M. indica*
Neem, Indian Lilac, Margosa Tree (Eng); Nim, Neem, Awadira d'Inde, Margousier, Azadirac (Fr); Azadira Hindi (Ar); Mwarobeini, Kohomba (Kisw); ________ ( )

Balanites aegyptiaca (L.) Del. (Balanitaceae)
syn. *Xemenia aegyptiaca, Agialida senegalensis, A. barteri, A. tombuctensis, Balanites ziziphoides*
Desert Date, Desert Torchwood, Egyptian Myrobolan (Eng); Myrobolan, Dattier du désert, Dattier sauvage (Fr); Heglig, Adoua (Hau); Mjunju, Mkonge (Kisw);
________( )

Bauhinia reticulata DC. (Leguminosae subfam. Caesalpinioideae)
syn. *Piliostigma reticulatum*
Camel's Foot Leaf Tree (Eng); Bauhenia, Semellier (Fr); Calgo (Hau); Mchikichi, Mchekeche, Mubamba Ngoma, Msegese, Mkoma, Msopo (Kisw);
________ ( )

Bombax costatum Pellegr. & Vuillet (Bombacaceae)
syn. *Bombax adrieui, B. houardii, B. vuilletii, B. buonopozensis*
Kapok, Silk Cotton Tree (Eng); Kapokier, Arbre à bourre (Fr); Kuria (Hau);
________ ( )

Borassus aethiopum C. Martius (Palmae subfam. Borassoidae)
syn. *Borassus flabellifera* var. *aethiopum*
African Fan Palm, Borassus Palm, Rhun Palm, Palmyra Palm (Eng); Rônier (Fr); Gigunia (Hau); Mvumo, Mtappa (Kisw); ________ ( )

Boscia angustifolia A. Rich. (Capparidaceae)
syn. *Boscia tenuifolia, B. patens*
Agahini (Hau); Chieh (Somali); Emejan (Turkana); Likwon (Pokot); ________ ( )

Boscia coriacea Pax
syn. *Boscia teitensis*
Mnafisi (Kisw); Edung (Turkana); Sorichon (Pokot); ________ ( )

Boscia minimifolia Chiov. ex desc.
Megag (Somali); ________ ( )

Boscia salicifolia Oliver
syn. *Boscia powellii*
Zouray (Hau); Mguruka (Kisw); ________ ( )

Boscia senegalensis (Pers.) Lam. ex Poiret
syn. *Podoria senegalensis, Boscia octandra*
Mokheit (Ar); Hansa, Dielo (Hau); ________ ( )

Butyrospermum paradoxum (Gaertn. f.) subsp. *parkii* (G. Don) Hepper (Sapotaceae)
syn. *Butyrospermum parkii, Vitellaria paradoxa, Bassia parkii*
Shea Butter Tree (Eng); Karité, Arbre à beurre (Fr); Lulu (Ar); Kandaya (Hau);
________ ( )

Cadaba farinosa Forssk. (Capparaceae)
syn. *Cadaba mombassana*
Suraya (Ar); Baggahi (Hau); Mvunja-vumo, Kibaazi Mwitu (Kisw); ________ ( )

Cadaba glandulosa Forsskal; ________ ( )

Caesalpinia decapetala (Roth) Alston (Leguminosae subfam. Caesalpinioideae);
________ ( )

Cajanus cajan (L.) Millsp. (Leguminosae subfam. Papilionoideae)
Pigeon Pea, Red Gram (Eng); Pois cajou (Fr); Mbaazi (Kisw); ________ ( )

Calliandra calothyrsus Meissn. (Leguminosae subfam. mimosiodeae)
syn. *Calliandra confusa, C. similis*; ________ ( )

Callitris glauca R. Br. (Cupressaceae)
syn. *Callitris columellaris, C. robusta*
White Callitris (Eng); ________ ( )

Calotropis procera Aiton f. (Asclepiadaceae)
Dead Sea Fruit, Apple of Sodom, Calotropis, Swallow-wort (Eng); Arbre à soie, Pomme de Sodome, Calotropis (Fr); Tumfafva (Hau); Mpamba Mwitu (Kisw);________ ( )

Capparis decidua (Forsskal) Edgew. (Capparaceae)
syn. *Capparis aphylla, C. sodata, Sodada decidua*
Capers (Eng); Tundub (Ar); ________ ( )

Carica papaya Linn. (Caricaceae)
Pawpaw, Papaya (Eng); Papayer (Fr): Mpapai (Kisw); ________ ( )

Cassia siamea Lam. (Leguminosae subfam. Caesalpinioideae)
syn. *Cassia florida, Sciacassia siamea*
Yellow Cassia, Iron Wood (Eng); Mjohoro, Mti-ulaya (Kisw); ________ ( )

Cassia spectabilis DC.
syn. *Cassia humboltiana, C. speciosa, Pseudocassia spectabilis*
Scented Shower, Calceolaria Cassia (Eng); ________ ( )

Casuarina cunninghamiana Miq. (Casuarinaceae)
Greek Oak, River She-oak (Eng); Mvinje (Kisw); ________ ( )

Casuarina equisetifolia J.R. & G. Forst.
syn. *Casuarina litorea*
Casuarina, She-oak, Whistling Pine (Eng); Filao (Fr); Mvinje (Kisw); ________ ( )

Casuarina glauca Sieb. ex Spreng.; ________ ( )

Cedrela serrata Royle (Meliaceae)
syn. *Toona serrata*
Mwerezi (Kisw); ________ ( )

Cedrela toona Roxb. ex Rottler & Willd.
syn. *Toona ciliata*
Toon, Burmese Cedar (Eng); Mwerezi (Kisw); ________ ( )

Celtis africana Burm. f. (Ulmaceae)
syn. *Celtis kraussiana, C. rhamnifolia, C. rhamricfolia*
African Elm, White Stinkwood (Eng); ________ ( )

Citrus limon (L.) Burm. (Rutaceae)
syn. *Citrus limona*
Rough Lemon (Eng); Citronnier, Limonier (Fr); Mlimau (Kisw); ________ ( )

Citrus sinensis (L.) Osbeck
Sweet Orange (Eng); Oranger (Fr); Mchungwa (Kisw); ________ ( )

Cocos nucifera L. (Palmae)
Coconut Palm (Eng); Cocotier (Fr); Mnazi (Kisw); ________ ( )

Colophospermum mopane (J. Kirk ex Benth.) J. Kirk ex J. Leonard (Leguminosae subfam. Caesalpinioideae)
syn. *Copaifera mopane*
Turpentine Tree, Mopane (Eng); ________ ( )

Combretum glutinosum Perrottet ex DC. (Combretaceae)
syn. *Combretum passargei, C. leonense*
Ratt (Fr); Taramnia (Hau); ________ ( )

Combretum molle R. Br. ex G. Don
Bush Willow (Eng); Wuyan Daho (Hau); Mlama (Kisw); ________ ( )

Combretum schumannii Engl. (Combretaceae)
syn. *Combretum macrostigmateum*
Mgurure, Mpera-mwitu (Kisw); ________ ( )

Commiphora africana (A. Rich.) Engl. (Burseraceae)
syn. *Heudelotia africana, Commiphora pilosa, C. calcicola, C. abyssinica, Balsam odendron africanum*
African Bdellium (Eng); Myrrhe africaine, Bdellium d'Afrique (Fr); Gafal (Ar); Ikitchi, Dashi (Hau); Mbambara, Mponda, Mkororo, Mtwitivi (Kisw); ________ ( )

Commiphora ellenbeckii Engl.
Myrrh Tree (Eng); Mkororo (Kisw); Melmel (Somali); ________ ( )

Cordeauxia edulis Hemsley (Leguminosae subfam. Caesalpinioideae)
Yeheb Nut (Eng); Yicib (Somali); ________ ( )

Cordia abyssinica R. Br. (Boraginaceae)
syn. *Cordia holstii, C. africana, C. ubanghensis*
Mukumari, Mringaringa (Kisw); ________ ( )

Cordia sinensis Lam.
Nyamate, Mkamasi (Kisw); ________ ( )

Crateva adansonii DC. (Capparidaceae)
syn. *Cratava religiosa*
Umm Brixesa (Ar); Gude (Hau); ________ ( )

Crotalaria ochroleuca G. Don (Leguminosae subfam. Papilionoideae)
Sunhemp (Eng); Marajea (Kisw): ________ ( )

Croton macrostachys Hochst. ex Del. (Euphorbiaceae)
syn. *Croton amabilis*
Koribe (Hau); Msinduzi (Kisw); ________ ( )

Croton megalocarpus Hutch.
syn. *Croton elliottianus*: ________ ( )

Cupressus lusitanica Mill. (Cupressaceae)
syn. *Cupressus benthamii, C. lindleyi, C. glauca*
Mexican Cypress (Eng); Cyprès (Fr); ________ ( )

Dalbergia melanoxylon Guill. & Perr. (Leguminosae subfam. Papilionoideae)
African Ebony, Senegal Ebony, African Blackwood, Zebra Wood (Eng); L'ébène du Sénégal, Dalbergia à coeur noir (Fr); Mpingo, Mugembe (Kisw); ________ ( )

Dalbergia sissoo Roxb. ex. DC.
Sissoo (Eng); ________ ( )

Daniellia oliveri (Rolfe) Hutch. & Dalz. (Leguminosae subfam. Caesalpinioideae)
syn. *Paradaniellia oliveri*
African Copaiba Balsam (Eng); Satan (Fr); Maje (Hau); ________ ( )

Delonix elata (L.) Gamble (Leguminosae subfam. Caesalpinioideae)
syn. *Poinciana elata*
Flame Tree, Flamboyant (Eng); ________ ( )

Delonix regia (Boj. ex Hook.) Raf.
syn. *Poinciana regia*
Flamboyant (Eng); Flamboyant (Fr); Msonobari, Mkakaya (Kisw); ________ ( )

Dichrostachys cinerea (L.) Wight & Arn. (Leguminosae subfam. Mimosoideae)
syn. *Dichrostachys glomerata, D. nutans, D. platycarpa, D. arborea, Mimosa glomerata, M. nutans, M. cinerea, Cailliea dichrostachys*
Chinese Lantern Tree, Sicklebush (Eng); Mkingiri (Kisw); ________ ( )

Diospyros mespiliformis Hochst. ex A. DC. (Ebenaceae)
syn. *Diospyros senegalensis*
West African Ebony, Monkey Guava, Jackal Berry (Eng); Ebenier, Ebenier de l'Afrique de l'ouest (Fr); Dundu (Hau); Mpweke (Kisw); ________ ( )

Dobera glabra (Salvadoraceae)
syn. *Dobera roxburghii*
Mpuka, Mswaki (Kisw); ________ ( )

Dombeya goetzenii K. Sch. (Sterculiaceae); ________ ( )

Dombeya rotundifolia (Hochst.) Planch.
syn. *Dombeya reticulata, D. verdoorn*
Wild Pear, Plum Blossom (Eng); Mtobwe (Kisw); ________ ( )

Dovyalis caffra (Hook. f. & Harv.) Warb. (Flacourtiaceae)
syn. *Aberia caffra*
Kei Apple, Wild Apricot (Eng); ________ ( )

Entada abyssinica A. Rich (Leguminosae subfam. Mimosoideae)
Mfwansiku (Kisw); ________ ( )

Erythrina abyssinica Lam. ex DC. (Leguminosae subfam. Papilionoideae)
syn. *Erythrina platyphylla, E. tomentosa*
Lucky Bean, Coral Tree, Red Hot Poker Tree (Eng); Mwamba-ngoma (Kisw); ________ ( )

Erythrina senegalensis DC.
Coral Flowers (Eng); Erythrine du Sénégal, Arbre corail (Fr); Madjirya (Hau);
( )

Eucalyptus camaldulensis Dehnh. (Myrtaceae)
syn. *Eucalyptus rostrata*
Red River Gum, River Gum (Eng); Mkaratusi (Kisw); ________ ( )

Eucalyptus cladocalyx F. Muell.
syn. *Eucalyptus corynocalyx*
Sugar Gum (Eng); Mkaratusi (Kisw); ________ ( )

Eucalyptus globulus Labill.
syn. *Eucalyptus maidenii*
Southern Blue Gum, Maiden's Gum (Eng); Mkaratusi (Kisw); ________ ( )

Eucalyptus microtheca F. Muell.
syn. *Eucalyptus coolabah*
Coolabah, Tiny Capsule Eucalyptus (Eng); Mkaratusi (Kisw); ________ ( )

Eucalyptus tereticornis SM.
syn. *Eucalyptus umbellata*
Forest Red Gum (Eng); Mkaratusi (Kisw); ________ ( )

Euphorbia balsamifera Ait. (Euphorbiaceae)
syn. *Euphorbia sepium, E. rogeri*
Balsam Spurge (Eng); Agoua (Hau); ________ ( )

Euphorbia tirucalli L.
syn. *Euphorbia mauritanica*
Finger Euphorbia, Milk Bush, Petrol Tree (Eng); Arbre de St. Sebastien (Fr); Mnayari, Utupu, Mchakaazi, Malangali, Mwasi (Kisw); ________ ( )

Ficus benjamina (Moraceae); ________ ( )

Ficus capensis Thunb.
syn. *Ficus mallatocarpa, F. sur, F. lichlensteinii, Sycomorus capensis*
Cape Fig, Bush Fig, Wild Fig, Broom Cluster Fig (Eng); Mkuyu, Mwangayo (Kisw);
________ ( )

Ficus natalensis Hochst.
Bark Cloth Fig (Eng); Arabi, Mlandege (Kisw); ________ ( )

Ficus sycomorus L.
syn. *Ficus gnaphalocarpa, F. damarensis, Sycomorus gnaphalocarpa*
Sycamore, Bush Fig (Eng); Gomeiz (Ar); Baoure (Hau); Mkuyu (Kisw); ________ ( )

Gleditsia triacanthos L. (Leguminosae subfam. Caesalpinioideae)
Honey Locust (Eng); Fevier (Fr); ________ ( )

Gliricidia sepium (Jacq.) Walp. (Leguminosae subfam. Papilionoideae)
syn. *Gliricidia maculata, Robinia sepium*; ________ ( )

Gmelina arborea Roxb. (Verbenaceae)
White Teak (Eng); ________ ( )

Grevillea robusta A. Cunn. ex R. Br. (Proteaceae)
syn. *Grevillea ambricata, G. peineta*
Silky Oak, Silver Oak (Eng); Msongoma (Kisw); ________ ( )

Grewia optiva Drummond ex Burret (Tiliaceae); ________ ( )

Grewia tenax (Forsskal) Fiori
syn. *Grewia betulifolia, G. populifolia*
Umm Ageda (Ar); Damak (Somali); ________ ( )

Guiera senegalensis J.F. Gmel. (Combretaceae)
Nger, Guiera du Sénégal (Fr); Sabara (Hau); ________ ( )

Hagenia abyssinica (Bruce) J.F. Gmel. (Rosaceae)
syn. *Brayera anthelmintica*; ________ ( )

Hyphaene coriacea Gaertner (Palmae)
Doum Palm (Eng); Mkoma, Mlala, Mkocke, Mnyaa (Kisw); ________ ( )

Hyphaene thebaica (L.) C. Martius
Doum Palm, Egyptian Gingerbread Palm (Eng); Doum (Fr); Goriba (Hau); Mkocke (Kisw); ________ ( )

Indigofera arrecta Hochst. ex A. Rich. (Leguminosae subfam. Papilionoideae)
Indigo (Eng); Indigotier (Fr); ________ ( )

Jacaranda mimosifolia D. Don (Bignoniaceae)
syn. *Jacaranda acutifolia*
Jacaranda (Eng); ________ ( )

Jatropha curcas L. (Euphorbiaceae)
Barbados Nut Tree, Physic Nut Tree (Eng); Pourghere, Pignon d'Inde, Médicinier Beni, Fève d'Enfer (Fr); Mbono (Kisw); ________ ( )

Jatropha diditar; ________ ( )

Juniperus procera Hochst. ex Endl. (Cupressaceae)
syn. *Juniperus abyssinica, Sabina procera*
African Pencil Cedar (Eng); Mwangati (Kisw); ________ ( )

Khaya nyasica Stamp ex Baker f. (Meliaceae)
Mkangazi (Kisw); ________ ( )

Khaya senegalensis (Desr.) A. Juss.
syn. *Swietenia senegalensis*
African Mahogany, Senegal Mahogany (Eng); Caïlcedrat, Acajou du Sénégal (Fr); Madadji (Hau); Mkangazi (Kisw); ________ ( )

Lannea acida A. Rich. (Anacardiaceae)
syn. *Odina acida*
Lannea acide (Fr); Farou (Hau); ________ ( )

Leucaena leucocephala (Lam.) De Wit (Leguminosae subfam. Mimosoideae)
syn. *Leucaena glauca, L. latisiliqua, L. salvadorensis, Acacia glauca, Mimosa glauca, M. leucocephala*
Lucaena (Eng); Ipil-Ipil (Tagalog: Philippines); Mbaazi (Kisw); ________ ( )

Maerua angolensis DC. (Capparaceae)
Kermut (Ar); Chichiwa (Hau); Mkuruka, Mtunguru (Kisw); ________ ( )

Maerua crassifolia Forssk.
syn. *Maerua rigida, M. senegalensis*
Sareh, Aouina (Ar); Jiga (Hau); Mutunguru, Mlala-mbuzi (Kisw); ________ ( )

Maesopsis eminii Engl. (Rhamnaceae)
syn. *Maesopsis berchemoides*
Nduga, Musizi, Muhumula (Kisw); ________ ( )

Mangifera indica L. (Anacardiaceae)
Mango (Eng); Manguier (Fr); Mwembe (Kisw); ________ ( )

Manihot glaziovii (Euphorbiaceae)
Manicoba Rubber, Ceara rubber, Tree Cassava (Eng); Manioc géant (Fr); Mpira (Kisw); ________ ( )

Markhamia platycalyx (Baker) Sprague (Bignoniaceae)
Mtalawanda (Kisw); ________ ( )

Melaleuca quinquenervia (Cav.) S.T. Blake (Myrtaceae)
syn. *Melaleuca leucadendron*
Cajeput Oil Tree, Broad-leafed Paper Bark, White Paper Bark (Eng); Cajeputier, Arbre à Goménol (Fr); ________ ( )

Melia azedarach L. (Meliaceae)
syn. *Melia orientalis*
Azedarach, Bead Tree, Persian Lilac, China Berry Tree, Syringa (Eng): Lilas de Perse (Fr); ________ ( )

Mitragyna inermis (Willd.) Kuntze (Rubiaceae)
syn. *Mitragyna africana, Nauclea africana, Uncaria inermis*
Coe Tree (Eng); Kabe, Guijeja (Hau); Ngato (Ar-Chad); ________ ( )

Mitragyna robrostipula (K. Schum.) Havil.
Mromberombe (Kisw); ________ ( )

Moringa oleifera Lam. (Moringaceae)
syn. *Moringa pterygosperma*
Horseradish Tree, Drumstick Tree (Eng); Ben ailé, Nevedie, Pois quenique (Fr); Ruwag (Ar); Zogall (Hau); Mzunze, Mrongo, Mlonge (Kisw); ________ ( )

Moringa stenopetala
Cabbage Tree (Eng); Mlonge (Kisw); ________ ( )

Morus alba L. (Moraceae)
syn. *Morus indica*
Mulberry (Eng); Mûrier blanc, Mûrier du ver à soie (Fr); Mforsadi (Kisw); ______ ( )

Olea capensis L. subsp. *macrocarpa* (C.H. WR.) Verdoorn (Oleaceae)
syn. *Olea hochstetteri, O. welwitschii, O. laurifolia, O. macrocarpa*
Elgon Olive, Ironwood (Eng); Loleondo, Musharagi (Kisw); __________ ( )

Oncoba spinosa Forsk.
Snuff-box Tree, Wild Rose (Eng); Mdara (Kisw); ________ ( )

Opuntia ficus-indica (L.) Miller var. *inermis* Weber (Cactaceae)
syn. *Opuntia inermis*
Prickly Pear Cactus (Eng); Mpungate (Kisw); ________ ( )

Parkia biglobosa (Jacq.) Benth. (Leguminosae subfam. Mimosoideae)
syn. *Mimosa biglobosa, Parkia africana*
Locus bean tree, Nere (Eng); Néré, Arbre à farine, Arbre à fauve, Nété, Caroubier africain, Mimosa pourpre (Fr); Dorowa (Hau); ________ ( )

Parkia clappertonia Keay.
syn. *Parkia oliveri*; ________ ( )

Parkinsonia aculeata L. (Leguminosae subfam. Papilionoideae)
Parkinsonia, Jerusalem Thorn, Hardbean (Eng); Genet épineux (Fr); ________ ( )

Passiflora edulis Sims (Passifloraceae)
Purple Passion Fruit (Eng); Fruit de la Passion, Passiflore, Grenadille (Fr);
_______( )

Persea americana Mill. (Lauraceae)
Avocado Pear (Eng); Avocatier (Fr); Mlangalanga, Mpea, Mwembemafuta, Mparachichi (Kisw); ________ ( )

Phoenix dactylifera L. (Palmae)
Date Palm (Eng); Palmier dattier (Fr); Dabino (Hau); Mtende (Kisw); ________ ( )

Phoenix reclinata Jacq. (Palmae)
Wild Date Palm (Eng); Palmier du Sénégal (Fr); Mkindu (Kisw); ________ ( )

Pinus caribaea Morelet (Pinaceae)
syn. *Pinus ellotti, P. hondurensis, P. bahamensis*
Caribbean Pine (Eng); Pin des Caraïbes (Fr); Msindano (Kisw); ________ ( )

Pinus halepensis Miller
Aleppo Pine, Jerusalem Pine (Eng); Pin d'Alep (Fr); Msindano (Kisw); ________ ( )

Pinus kesya Royle ex Gord.
syn. *Pinus khasya, P. insularis*
Benguet Pine (Eng); Msindano (Kisw); ________ ( )

Pinus pinaster Ait.
syn. *Pinus hamiltonii*
Cluster Pine (Eng); Msindano (Kisw); ________ ( )

Pithecellobium dulce (Roxb.) Benth. (Leguminosae subfam. Mimosoideae)
syn. *Mimosa dulcis, Inga dulcis*
Manila Tamarind, Madras Thorn (Eng); Mchongoma (Kisw); ________ ( )

Podocarpus henkelii Stapf ex Dallimore & A.B. Jacks (Podocarpaceae)
syn. *Podocarpus thunbergii* var. *falcata*
Podo, Yellow Wood (Eng); ________ ( )

Polyscias fulva (Hiern) Harms (Araliaceae)
syn. *Polyscias ferruginea*
________ ( )

Populus euphratica Olivier (Salicaceae)
syn. *Populus diversifolia, P. ariana, P. bonnetiana*
Euphrates Poplar, Indian Poplar (Eng); ________ ( )

Prosopis africana (Guill. & Perr.) Taub. (Leguminosae subfam. Mimosoideae)
syn. *Prosopis oblonga, P. lanceolata*
Kiriya (Hau); ________ ( )

Prosopis arborea; ________ ( )

Prosopis chilensis (Molina) Stuntz
syn. *Ceratonia chilensis*
Mesquite (Eng); ________ ( )

Prosopis cineraria (L.) Druce
syn. *Prosopis spicigera, Mimosa cineraria*
Jand (Hindi); Ghaf (Ar); ________ ( )

Prosopis juliflora (SW) DC.
syn. *Mimosa juliflora*
Mesquite (Eng); ________ ( )

Prunus africana (Hook. f.) Kalkm. (Rosaceae)
syn. *Pygeum africanum*
Red Stinkwood (Eng); Mkomahoya (Kisw); ________ ( )

Prunus serotina Ehrh.
Mexican Sherry, Black Cherry (Eng); ________ ( )

Psidium guajava L. (Myrtaceae)
syn. *Psidium guianense*
Common Guava (Eng); Goyavier (Fr); Mpera (Kisw); ________ ( )

Pterocarpus angolensis DC. Poir. (Leguminosae subfam. Papilionoideae)
syn. *Pterocarpus erinaceus, P. echinatus, P. bussei*
African Gum, Kino Tree, African Rosewood, African Teak, Cornwood, Gambia Gum, Lancewood, Molompi Wood Tree, Senegal Rosewood (Eng); Vène, Bois de Sang Viène, Kino Vène, Ptérocarpe hérissé, Santal du Sénégal, Palissandre du Sénégal, Santal rouge d'Afrique, Olivier du Sénégal (Fr); Madobiya (Hau); Mninga, Mtumbati, Mhagata (Kisw); ________ ( )

Pterocarpus lucens Lepr. ex Guill. & Perr.
syn. *Pterocarpus abyssinicus, P. simplicifolius*
Taraya (Ar); ________ ( )

Rauwolfia caffra Sond. (Apocynaceae)
syn. *Rauwolfia goetzei, R. inebrians, R. natalensis, R. obliquinervis*
Quinine Tree (Eng); Mwembe Mwitu, Msesewe (Kisw); ________ ( )

Ricinodendron rautanenii Schinz (Euphorbiaceae)
Mongongo Tree (Eng); Muawa (Kisw); ________ ( )

Ricinus communis L. (Euphorbiaceae)
Castor Oil Plant (Eng); Ricin (Fr); Mbarika, Mbono (Kisw); ________ ( )

Robinia pseudoacacia L. (Leguminosae subfam. Papilionoideae)
Black Locust, False Acacia (Eng); Robinier faux acacia (Fr); ________ ( )

Salvadora persica L. (Salvadoraceae)
Mustard Tree, Toothbrush Tree (Eng); Araka (Ar); Kalahia (Hau); Mswaki (Kisw);
________( )

Schinus molle L. (Anacardiaceae)
Pepper Tree (Eng); Mpilipili (Kisw); ________ ( )

Sclerocarya birrea (A. Rich.) Hochst. ssp. *caffra* (Sond.) Kokwaro (Anacardiaceae)
syn. *Poupartia birrea, Spondidas birrea, Sclerocarya caffra*
Prunier du Sahel, Poupartia (Fr); Dania (Hau); Mng'ongo, Mongo, Marula (Kisw);
________ ( )

Sesbania grandiflora L.) Poir. (Leguminosae subfam. Papilionoideae)
syn. *Agati grandiflora*; ________ ( )

Sesbania sesban (L.) Merr. var. *sesban*
syn. *Sesbania aegyptiaca*
Egyptian Rattle Pod, Sesban (Eng); ________ ( )

Simmondsia chinensis (Link) C. Schneider (Buxaceae)
syn. *Simmondsia californica*
Jojoba (Eng); ________ ( )

Sterculia setigera Del. (Sterculiaceae)
syn. *Sterculia cinerea, S. tomentosa*
Sterculia (Eng); Mbep, Gommier mbep, Platare du Sénégal (Fr); Kukuki (Hau);
________( )

Syzygium cuminii (L.) Skeels (Myrtaceae)
syn. *Eugenia jambolan*
Black Plum (Eng); Jambolana, Mzambarau (Kisw); ________ ( )

Tamarindus indica L. (Leguminosae subfam. Caesalpinioideae)
Tamarind, Indian Date (Eng); Tamarinier (Fr); Tamar al Hind (Ar); Tsamiya (Hau);
Msisi, Mkwaju (Kisw); ________ ( )

Tamarix aphylla (L.) Karsten (Tamaricaceae)
syn. *Tamarix articulata, T. orientalis*
Salt Cedar, Tamarisk (Eng); ________ ( )

Tectona grandis L. f. (Verbenaceae)
Teak (Eng); Msaji (Kisw); ________ ( )

Terminalia brownii Fresen. (Combretaceae)
Mbarao, Mbambaro (Kisw); ________ ( )

Terminalia catappa L.
Indian Almond, Bastard Almond (Eng); Badamier, Amandier de Gambie, Myrobalanier (Fr); Mkungu (Kisw); ________ ( )

Terminalia macroptera Guill. & Perr.
syn. *Terminalia chevalieri, T. suberosa, T. adamanensis, T. elliotti, T. dawei*
________ ( )

Trema orientalis (L.) Blume (Ulmaceae)
syn. *Trema guineensis, T. bracteolata, Celtis orientalis*
Pigeon Wood (Eng); Mpesi, Msasa, Msinga (Kisw); ________ ( )

Vernonia amygdalina Del. (Compositae)
Bitterleaf (Eng); ________ ( )

Vitex doniana Sweet (Verbenaceae)
syn. *Vitex cuneata, V. chariensis, V. cienkowskii, V. paludosa*
Black Plum (Eng); Dumnjaa (Hau); Mfufu, Mfudu (Kisw); ________ ( )

Ximenia americana L. (Oleaceae)
syn. *Ximenia americana* var. *mircophylla, X. lauriana*
Wild Plum, Sour Plum, Wild Olive, Wild Lime (Eng); Citronnier de mer, Prunier de mer (Fr); Mpingi, Mtumbwitumbwi (Kisw); ________ ( )

Ziziphus mauritiana Lam. (Rhamnaceae)
syn. *Ziziphus jujuba, Z. orthacantha, Z. mauritiaca*
Jujube, Chinese Date, Indian Plum (Eng); Jujubier (Fr); Nabag (Ar); Magarya (Hau); Mkunazi, Jujube (Kisw); ________ ( )

Ziziphus mucronata Willd.
syn. *Ziziphus mitis*
Cat Thorn, Buffalo Thorn (Eng); Jujubier de la hyène (Fr); Mgagawe, Mkunazi Mwitu (Kisw); ________ ( )

APPENDIX III: GUIDELINES AND KEY QUESTIONS FOR INTERVIEWS

PART 1. SAMPLE QUESTION-ANSWER SHEET

Agroforestry workers may find it useful to prepare a set of answer sheets and sketches to summarize the information they already have about an area before they begin a field project. The same format can be used again to record information obtained on the first field visits, allowing workers to compare their images and ideas **before** and **after** taking a fresh look at the landscape, people and land-use systems in a particular place. The sample information summaries and exercises provided here may serve as a guide or reminder for planning and recording these first steps in the overall survey. Researchers and extensionists may also want to keep their answer sheets and sketches for future comparison of their own early impressions with the information provided later by the local community.

1. Name of Site/Area ____________________

2. Boundaries of the Site ____________________

2a. What are the site boundaries for purposes of this work?

____

____

____

____

2b. Sketch these on a separate sheet.

3. Group(s)

Who lives here, works here or uses resources in this area? ____________________

____

____

____

____

4. Climate and Ecological Zone ____________________

rainfall (mm/yr) ____________ elevation (m) ____________

temperature extremes ____________ minimum ____________ maximum

seasons ____________________

____

(for example, number and timing of planting seasons per year)

5. Natural Hazards/Problems

droughts ________ floods ________ windstorms ________

sandstorms ________ landslides ________ mudslides ________

extremes of temperature (explain) ____________________

fires ________ severe erosion ________ other ________

6. Local Terrain

Sketch (on a separate sheet) the local terrain and note the major variations in slope and the locations of rivers, streams, lakes, ponds, springs, swamps, seasonally flooded areas, gullies and severely eroded or degraded lands.

7. Soils

7a. List the major soil types in the area.

__________ __________ __________ __________

__________ __________ __________ __________

7b. Sketch the location (distribution) of these soil types on a separate sheet.

8. Vegetation and Land Use

8a. List the major classes of vegetation and land use. For example, you might find the following land-use and land-cover categories:

Land Cover	*Land Use*
forest	conservation reserves
woodland	gathering/collecting areas
savannah (trees over grassland)	grazing and browsing lands
open grassland	croplands
perennial crops	gardens
annual crops	homesteads
bare soil	public markets and meeting places

These categories will vary in different places, depending on the types of cover and the range of land uses. For instance, there may be two kinds of woodland—open grazed woodland and dense forest protected as a conservation area. In the same area, there might be several kinds of perennial crop cover on cultivated lands, including banana plantations, citrus orchards, multistorey home gardens and timber trees over coffee or tea. List the land-cover and land-use categories which best describe the range of conditions in the area:

Land Cover	*Land Use*
__________	__________
__________	__________
__________	__________
__________	__________
__________	__________
__________	__________
__________	__________
__________	__________
__________	__________

8b. On a separate sheet, sketch the distribution of these land-cover/land-use types in the local landscape.

8c. List some of the most **common** plants in the area:

crops	*grasses*	*shrubs*	*trees*

8d. List some of the most important plants:

crops	*grasses*	*shrubs*	*trees*

9. Settlements

9a. Describe below and sketch on a separate sheet the type(s) of settlement pattern, plus the roads, paths and waterways and their role in the larger landscape.

9b. Note the locations of waterpoints and crossings for people and animals and fill in the names of these places on your sketch from 9a.

10. Landscape Relationships

10a. Compare all the sketches and note any striking or important relationships between slope, soil, water, vegetation type, land use and settlements. For example, do settlements occur mainly along the slopes of hills, on hilltops, along rivers or in association with some other natural feature of the landscape?

10b. You may also find it convenient to note these related features (10a) on the original or separate sketches.

10c. For each land-use/land-cover type and other landscape feature, note who makes use of the place and/or the vegetation, who works there and who owns or controls the place and the resources on it.

11. Production or Resource-Management Problems

What do you think are the major resource-management and production problems in the area? Associated with what types of land use and in which places? Which group(s) of land users are involved? Cause? Effect?

Problem	*Land use*	*Place*	*Land-user group*
______	______	______	______
			______
______	______	______	______
			______
______	______	______	______
			______

12. Trees and Shrubs in the Landscape

12a. Where do trees and shrubs occur in the landscape?

_____ Home compound	_____ Property boundaries
_____ Cropland	_____ Fences
_____ Fallowland	_____ Roads, trails
_____ Pastures	_____ River banks
_____ Rangeland	_____ Gullies
_____ Woodlots	_____ Canals
_____ Woodlands, forests	_____ Waterholes

12b. Are the trees and shrubs in blocks, clumps, lines, dispersed or isolated?

____

____

____

____

12c. Draw a few sketches of how trees and shrubs fit into the landscape.

12d. Based on your own first impression, what are the best places to add or introduce trees?

_____ Home compound	_____ Property boundaries
_____ Cropland	_____ Fences
_____ Fallowland	_____ Roads, trails
_____ Pastures	_____ River banks
_____ Rangeland	_____ Gullies
_____ Woodlots	_____ Canals
_____ Woodlands, forests	_____ Waterholes

Why? ____

For each of the places noted above, who **controls** the place, the plants and their products and who **uses** the place, plants and their products?

Place	*Who Controls*	*Who Uses*	*Whose Labour Input*

Plants	*Who Controls*	*Who Uses*	*Whose Labour Input*

Products	*Who Controls*	*Who Uses*	*Whose Labour Input*

12e. Also based on your own best guess, what species? In what arrangement? Why?

Place in landscape	*Species*	*Arrangement*	*Reason/Purpose*

Would the new practices listed above change the terms of control over the place, the plants on site or their products, as noted in 12d above? Would it affect the terms of access by various land users?

Try to repeat this exercise during or just after the first general field survey, to compare what you thought initially with what you later saw and learned. What were the major differences between the two sets of sketches and answers?

Then, compare both these results with the way people in the community see their own settlements, the surrounding environment, local production systems and the place of trees in the landscape. Note the differences between your original impressions, your observations and the views of various groups in the community. As part of an evaluation, you might compare answers in 1 or 2 years to see what you and each group learn from your own experiences and from each other.

PART 2. GROUP INTERVIEWS ON GENERAL TOPICS

The main topics of discussion, which apply almost anywhere, are (1) an introduction to the people, their history, their place and how they see and use it; (2) current practices in the management of crops, livestock, land and water; (3) needs and future plans of the people; (4) land-use problems and their solutions; (5) land-use change; (6) specialized knowledge about management and use of land, water, plants and animals; and (7) future directions for agroforestry and other sustainable production systems.

To cover all of the topics and questions listed below would take more than a single visit with a group. However, you may decide to discuss each topic briefly and then follow up with one or two more group interviews plus individual interviews on specific topics.

Alternatively, you may choose to divide group meetings into three or more fairly detailed discussions on separate topics: introduction, land-use systems and local practices and specific sessions on trees, shrubs, grasses and other plants. The sample questions and discussion points presented here should provide some ideas about how to learn with local people about the existing and potential roles of trees and agroforestry practices in local land-use systems.

The questions can be followed closely, although they are best used as starting points to encourage people to describe and discuss the various topics. Check occasionally for variations in practice within the group, by asking for a show of hands, voices or nodding. Be careful not to let one person or a few people dominate the discussion. If necessary, call on specific people occasionally or ask for the opinion of the whole group through voting.

1. Introduction

Once you are able to meet the group, introduce yourself and your reasons for seeking local knowledge and opinions about land-use systems. For example, you might explain that you are getting to know the place in order to do a better job of rural development, together with their help. If you know the place, and are known, it may still help to explain that this exercise will broaden your work to serve people's interests more fully. If you don't know the group, it is also useful to ask them to explain briefly who they are as an introduction.

On a summary sheet, note the information required to locate, identify and describe the people present, including how they reflect the characteristics of the group as a whole. These questions may also be answered by someone who knows the group, rather than quizzing the group about all of the details.

1.1. Site name ____

1.2. Site description (landforms, land use, vegetation, settlement, proximity to roads, streams, markets, landmarks)

____

____

1.3. Group name or description ______________________________

____

1.4. If formal group: purpose, history, total number of members, type of people who are members

Purpose ______________________________

History ______________________________

Total number of members ______________________________

Description of members (group as a whole, sub-groups). For example, note the proportion of men/women, farmers/herders, wealthy/poor, valley dwellers/hill slope farmers or any other distinctions that would help to describe the members.

____

____

1.5. Description of those present at the discussion

Number present ______________________________

Composition of group present ______________________________

Comparison to group as a whole ______________________________

1.6. Where do the people in the interview group live? (Note on a sketch map of the area.) ______________________________

____

1.7. Where are they (and their parents) from originally?

Group members ______________________________

Elder generation ______________________________

1.8. If they or their parents came from somewhere outside the area:

Where was it? ______________________________

What was it like?

Land ______________________________

____

Climate ______________________________

____

Land use ______________________________

____

Settlements ______________________________

____

When and why did they leave and move to this place? ______________________________

____

____

2. Livelihoods and Land Use

2.1. Ask about the types of land use and production systems practised locally. For example, these systems might focus on: commercial citrus groves; dairying; charcoal or timber production; mixed farming with crops and livestock; seasonal migratory herding of sheep, goats and cattle; group or communal ranching; commercial ranching; or subsistence farming.

Land-use/production systems

2.2. What do people do to make a living? List specific occupations and activities of men, women and children, both paid and unpaid.

2.3. Do some people work for wages? Do they earn wages locally or outside the area? What jobs do they do? Where? For whom?

Type of job	*Who: number and type of people (men, women, children?)*	*Where*

2.4. Are there some things that only a few specialists do? Are any of these specialists present? If so, note for possible follow-up interviews.

2.5. Among those present, what kinds of production system are they involved in? Read back the list of different land-use systems and activities, including working as hired labour in particular systems. Ask for a show of hands to get a sense of whether everyone participates in all activities or whether people are working in different occupations or different production systems. For example, are there separate groups of herders and farmers? Or goat herders and cattle herders? Or cash-crop farmers and subsistence farmers? Or charcoal makers and timber harvesters? Or small-scale commercial/subsistence farmers and large-scale commercial/subsistence farmers? Or absentee men farm owners and women farm managers?

Land-use system	*Occupation*	*Wage or subsistence*	*Number of people men/women*

2.6. Besides occupation, what other categories divide land-user groups in the community? Check the lists in Chapter 2 or ask the group what characteristics they would use to categorize individuals and groupings of people in the community (by age, sex, wealth, ethnic group, place of origin, current location, religion?).

2.7. What kinds of groupings exist and which are the most important with respect to land use (family, extended family, clan, club, association, co-operative, collective)? Do men and women belong to the same kinds of grouping; does each grouping have the same relative importance for men and for women?

Grouping	*Type of members*	*Relative importance*	*Importance for land use*	*Authority over land, resources, products*

3. Land, Land-Use History and Natural Resources

The objective of this exercise is to learn the local names of the different categories of natural resources and the logic which divides one from the other. This will be important for any discussions of land use problems and possible changes. While it will rarely be possible or practical to document these systems in detail, it is essential to be able to discuss resource-management and production problems and agroforestry practices in terms of specific classes of land and other resources which have some meaning to local people.

3.1. How many different types of 'lands' are there and what are they called?

Land type	*Characteristics*	*Example (place)*

If you are in a position to see the surrounding area, ask people to point out or to sketch where these various types of land are, or ask who might be able to help you later in a separate mapping exercise. In either case, draw a rough map labelled with the local names for each land type and keep it for reference in further discussions and surveys.

Also be sure to include both men and women (or other important sub-groups) in the exercise. In one or two cases, try to obtain separate maps or descriptions from different groups to find out to what extent they see and evaluate their surroundings differently. This may affect their participation in future agroforestry activities.

3.2. Do people define land types by vegetation, soils, landforms, suitability for different land uses, current land use, a combination of these or other characteristics?

3.3. Aside from general categories of land type, how do people categorize soils? Landforms? Vegetation? Land uses?

Soils	*Landforms*	*Vegetation*	*Land uses*	*Other**

*For example, herders may distinguish 10 classes of grazing or browse or women in dry areas may distinguish several classes of water source or water quality.

During the group discussion or in follow-up discussions with a few group members, make sketch maps of each of the resource classifications above. Try to draw these maps at roughly the same scale as your general land-type map to allow comparisons.

3.4. In addition to using land, people usually judge the value of different kinds of land, either in general or for particular uses. Ask people to point out the cropped fields, the grazing lands, the gathering grounds, the settlements. Which are the best, the acceptable and worst places for each use and why? Where are the water sources? Which are the best, and why? Where are the main roads and trails?

Ask people to point out or to name places in each value category and note these for future mapping and field visits. Ask the group to identify appropriate individuals for follow-up interviews on evaluation of land and water resources for particular uses.

In follow-up individual interviews, it should also be possible to determine if land and resource categories refer to permanent properties, present conditions or both. For instance, people often use several categories to describe the different stages of the cropping cycle over time: a newly planted field, a field that has been cropped for several years, a 2-year fallow and a 4-year fallow might all have specific names rather than being called simply cropland or fallow. The same is true for different kinds of pasture and woodland, including categories that refer to degree of degradation, intensity of use or stages in a rotational cycle.

4. Land-Use History and the Changing Condition of Natural Resources

4.1. What was the area like when the eldest members of the group were young, or when they first settled there?

Land____________________

Water____________________

Soil____________________

Vegetation____________________

Wildlife____________________

Land use____________________

Local economy____________________

Erosion features/conditions____________________

____

By pointing to reference sites in the existing landscape, plus verbal descriptions and people's own drawings (on paper, chalkboard or ground), sketch this remembered landscape. Note roughly the year or the time period that it represents. Be sure that the descriptions and sketches include such features as forests, grasslands, croplands, water resources, roads and settlements.

4.2. What major changes in land use have taken place? List these, ask where and when they took place and note them on the sketches.

Original land use	*Change*	*Where*	*When*

4.3. What are the major changes in the condition of natural resources?

Resource	*Change: degradation/improvement*	*Where*	*When*

Note the locations of these changes on sketch maps.

4.4. What measures have people taken to stop degradation of resources or to improve the condition of soil, water, natural vegetation or wildlife?

Resource	*Problem*	*Control or improvement practice*

Ask the group to describe these practices and to note the location of specific examples for later field visits and for reference in future discussions. Note these on sketch maps.

4.5. Have government or outside organizations introduced any special conservation practices? Which practices? When? What were the results? How did people feel about these practices?

Practice	*Organization*	*Where*	*When*	*Results*

5. Crop and Livestock Production

Usually one of the easiest topics to discuss, crop and livestock production may be the first information obtained as the group describes their land-use systems and land-use history. If so, use this section as a checklist to fill in the information gaps that remain. You may also choose to begin with this section, just after your introduction to the group. Most of the questions can apply to either crops or livestock, although a few separate examples are included.

5.1. What crops are grown? Ask people to name all of the crops and compile a list. Then read back the list and ask which are for home use, for sale or both and note these on the list. Also get a rough estimate of how many members, among those present, grow each of these crops.

Crop/variety	*Sale or home use*	*Who grows (works on) it group/number (men/women/children?)*	*Who owns it group/number (men/women/children?)*	*Where*

5.2. Which crops are grown together or in regular rotations? List these and note whether they are combined at the same time or follow in rotation.

Crop combination	*Who grows it group/number (men/women/children?)*	*Where*

5.3. Do most people have one field for cropping, or more? If more, why? What size are most people's cropped fields? Notice how people describe the plot sizes and areas. Try to learn and use their terms in future discussions. Ask how many plots most people have. Ask for a show of hands to indicate numbers and sizes of plots if this is not too sensitive an issue.

Number of plots	*Total area of plots*	*Numbers/types of people (men/women/other sub-groups?)*

5.4. Follow up with sketches of a few people's landholdings or land used, including outlying plots. Choose people representative of different groups, such as rich/poor, men-/women-headed households etc. Sketch the cropping system on individual plots as well as the locations and types of plot within the landscape. For those same households, note on the sketches which places are controlled, used and maintained by men, women and children respectively.

5.5. Do people clear forest, range or bush fallow for cultivation? If so, how do they clear the land? How do people prepare land that has already been cleared? Hoe, digging stick? Plough? If ploughs are used, are they drawn by oxen, donkeys, horses, cows? Does anyone use small motorized tillers? Tractors?

Clearing methods	*Tillage methods*	*Animals used*	*Who clears, tills*

5.6. Do people crop the same fields permanently? How many crops per year? If not permanent, how many years before they change, and why? What happens to the fallow? When do they come back to the original cropland again, and why? How much variation is there in the group? Obtain one or two responses, then ask how many people follow a different timetable or do something differently? Obtain enough explanations about cropping practice to know how many distinct situations there are with respect to crop rotation and fallow cycles.

____

____

____

____

5.7. Do people use any form of irrigation or water harvesting? Do they use any soil- or water-conservation structures, such as terraces, bunds or drainage ways, in their croplands? Do they use any temporary structures, such as tied ridges or mounds?

Practices	*Where*	*Who builds/maintains*	*Who owns*

5.8. Gather enough information about different kinds of activity to draw a rough cropping calendar (see example on page 269). Who does the work? Who helps? When? Who decides when?

Activity	*Who works/decides*	*Where*	*When*

5.9. All the same questions can apply to livestock production and herding activities or their combination with crop production. Maps and sketches of grazing lands are also useful. For example, what kinds of livestock do people have? Why? Who has what kind of animals?

What animals? species/breed?	*What products? or services?*	*Sale or home use*	*Who owns? groups/numbers*	*Who manages? groups/number*

What combinations of animals are owned by those present	*How many of each species do those present own?*	*How many managed? (own/others' animals)*

Who owns animal products? (for sale and home use)	*Who decides management including sale and purchase?*

6. Trees and Other Plants

The past and present use and knowledge of trees in farming and herding communities can provide a solid basis for the future development or improvement of agroforestry practices. The questions below provide a basic list of species, uses, preferences and future possibilities. These questions could be included in a more general interview. For more detailed information on plants, see the special-topic interview in Part 3.

6.1. How do people (and their animals) use trees and shrubs? List the uses, and then read the list to the group. Ask if any have been left out. If so, note any additions. Ask the group to list particular plants for each use and note the plants by name* on a copy of the blank species list given as Table 3 in Appendix I. If the list becomes too long, you may have to limit it to the best and most important species for each use and compile a more complete list in follow-up interviews.

If people do not seem comfortable listing plants by use, try to get a list of all the plants they know and use, and then list all the uses for each plant. However, for most purposes, it is usually more helpful to list plant species under each use, rather than listing uses under each species.

6.2. Note how many people say that they use each plant for a specific purpose. Ask for a show of hands. Are they men, women, children, mixed? Are some plants (medicinal, cosmetic, ceremonial) reserved for use by specialists? Is the use of any of these plants otherwise restricted?

*It is important to learn the local names of common plants. For your own use, prepare a list in a small notebook with spaces for local and Latin names and fill it in during the course of your work.

Use	*Species*	*Who uses*

Ask the same questions about shrubs, grasses and other plants. List the uses and the plant names, following the same procedure. Use copies of Tables 3 and 4 in Appendix I or any other convenient form to record the group's replies.

6.3. Now return to the species lists and ask people to indicate the relative importance of the various plants. For example, read the list of choices, then ask for a show of hands for the plant which rates highest, and read each name again for a vote. Which is the favourite for a particular use? Which are most important? Which wild plants could be domesticated? Ask 'why' throughout the exercise and make notes about famine foods, dry-season fodder, special qualities of products etc.

Use	*Preferred species*	*Important species*	*Priority species for domestication*	*Who uses/controls (numbers/group)*

6.4. To apply all this information to agroforestry and related practices, you also need to know **where** people are willing to plant or manage trees. Where have trees and shrubs been maintained or planted in the past? In large range and forest tracts? In blocks? In lines? Dispersed? As single trees? Where?

Plant name	*Planted or maintained*	*Where (land type, site)*	*Arrangement*	*Who uses/controls? (numbers/group)*

6.5. Have there been any changes in the kinds of space that trees occupy in the landscape? Are there any spaces where trees or shrubs might be introduced or increased where they have not been kept before? Ask people to name the best places. Note differences between men's and women's preferences and between those of different age and economic groups.

6.6. Keep in mind which places in the landscape have not been mentioned. When the group has finished naming the likely places for trees, ask them about some of the other places, and especially about their reasons for not putting trees there. This can help you to understand whether trees and shrubs could grow in some of these places if they were specially managed or whether they simply could not.

____

____

____

____

7. Discussing Problems Associated with the Land-Use System

People can almost always name a few key problems they would like to resolve. One way to focus the discussion is to start with basic needs and production.

7.1. What are the major problems in household production systems?
Discuss these in terms of domestic needs, such as food, water, fuel, shelter, cash, investment, inheritance, raw materials for crafts and resources to meet social obligations. Make a list of several problems and ask for a show of hands to indicate which ranks first and second in terms of importance. Do different groups have different problems? Or just different priorities? Note who feels most strongly about the problem—does it depend on age, sex or economic position?

Problem	*Where*	*Who*

7.2. What are the major problems in production systems at the community level? Are there resource-management, supply or production problems specific to particular places in the landscape or to particular types of land? Cropland? Grazing land? Gathering grounds and woodlands? Settlements? Water sources? Drainage features? Roads and trails? Public markets and meeting places? Again, make a list, discuss it and ask people to vote on the relative importance of problems faced by the community as a whole. Note if there are clear divisions between sub-groups (for instance, if men say the most important problem is fodder, whereas women say it is fuelwood).

Problem	*Where*	*Who*

7.3. What have people done about these problems in the past? What succeeded, what failed, and why?

Problem	*Previous response*	*Results*

8. Parting Questions

Before ending a general group interview, review the key topics of discussion and choose a few points which require decisions or statements of preference or priorities. Put these questions to the group again and ask them to think over and discuss these with their families, friends or other groups before the next meeting. Set a time for the next meeting, allowing 1 to 2 weeks for people to reflect upon the questions.

PART 3. SPECIALIZED INTERVIEWS

1. Overview

Field workers may wish to conduct specialized interviews with individuals or small groups. In either case, the purpose is to discuss a specific topic in detail. This should be viewed more as a joint working session than as an interview. Usually the session begins with lists and basic information and gradually develops into a fuller explanation of complex knowledge and practice. In some cases, these sessions also lead to discussions of controversial issues and alternatives.

The sample discussion summarized here focusses specifically on information about plants. The questions cover types of plant, their uses and their ecology, including site requirements and management. The discussion expands toward the end to include such issues as domestication of plants and their niches in the larger landscape.

The same kind of in-depth, single-topic discussion could also focus on soil conservation, water management or another subject. The example of trees and their uses is given because it is usually crucial for agroforestry projects. However, similar interviews could yield valuable specialized information and stimulate discussion on a wide range of natural resources or production systems. For example, this process might be used to discuss charcoal making, tree-product processing for food and condiments, tree-product marketing, the use and preparation of medicinal plants, management of home gardens, soil conservation practices or livestock management.

2. An example

The most important information for most agroforestry projects will be: What trees and shrubs do people already use? Who uses what? Who controls access to these plants? Where are they located? How are they managed? How are they used? Why these plants and not others? Some knowledge of changes in all of the above can also let you know if favourite plants are disappearing, if access is becoming difficult, if old skills or sharing arrangements are being eroded, or if markets and local preferences are changing. These can point out critical areas of concern where there is a widening gap between what woody plants people have and what they would like to have or need. Where such gaps are growing, people may be motivated to move quickly to recover past conditions or to introduce and develop new agroforestry production systems that provide the same products and services. In either case, it is essential to pay careful attention to past practice, the present state of resources and current uses and preferences with respect to particular plant species.

To conduct a specialized interview, inform people in advance about the topic and allow sufficient time for the meeting—up to 2 hours. Introduce yourself, if necessary, and explain your interest in people's use and knowledge of plants, especially wild plants and trees. Begin, as in more general interviews, by asking about how people use trees. Which species do they use for which purpose? Which species do they prefer for each use? See Tables 3 and 4 of Appendix I for summary sheets.

Then, for each species named ask: Is it abundant, just enough, hard to find or in short supply? Where is it found? In what kinds of soil? On what kinds of landform (slopes, ridges, plains, stream banks)? Does it commonly occur in association with other plants? Which ones? Is it wild, managed or planted? What is the land use where it is found? Who knows this plant, its uses and its habitat well?

Who has information on special topics such as soils and site requirements of plants, plant pests and diseases or pesticidal or medicinal plants? If this is a group discussion, identify individuals for follow-up interviews. If talking with one person, compile a list of suggested experts for further interviews.

What do people need and want that trees could provide? List their initial answers, then check the following: Are they interested in cash income, products for home use or both? Do they want trees or shrubs to protect or enrich the soil or to help conserve water? Do they want trees to help protect other plants in some way? Is there any experience or interest in trees as a form of savings, investment or inheritance? Are people interested in trees to define and decorate different spaces in the landscape? Is anyone interested in rights to trees and their products, or using trees to affect land rights?

Specifically, what do people want from trees and shrubs? Ask what products and services people would like from new planting or management practices or from new species. Use a copy of Table 3 in Appendix I to note how many people want each of the products and services listed.

Which specific trees do they want? Which shrubs, grasses and other plants? People may identify species they already know, like and want to domesticate, plant more of, put into new places and combinations or use in new ways. Even in cases where these favourite species grow slowly, or do not fit well into proposed new sites, combinations or uses, the reasons for choosing them may provide important clues as to which other trees—including exotics—might fit peoples' needs and preferences. For uses and species already noted, are the preferences widely shared or specific to particular groups (based on sex, age, family, ethnic group etc.)? If there are differences among groups, it may be best to compile separate lists.

Assuming there is some knowledge about the trees themselves and about which ones people want, what local resources are there for raising, managing, and planting trees, shrubs and grasses for agroforestry systems? What do people know about growing and managing trees, shrubs and grasses? How much experience have they had with managing these plants? Which kinds and in which environments?

Have they ever transplanted these species from the wild? Or collected or treated seeds? Or planted seeds directly at a site or in seed beds? Have they ever raised seedlings in a nursery? Have they ever made and planted cuttings? Have they ever planted seedlings (their own, purchased or gifts)? Do people have any experience with grafting, layering or other horticultural techniques?

For each type of knowledge or experience, is it widely shared by the people of the community or confined to a particular group (based on sex, age, family, ethnic group) or to a few specialists? Or is it simply not part of local experience?

What resources are available at the community or district level? Are there government, private or group nurseries and seed sources? If so, how far away are they, how accessible and to what extent do people already use these facilities? If they do not use them, why not? What species are available in these seed centres or nurseries? What do they most commonly distribute—seeds, seedlings or cuttings? Are these facilities able to stock new species, based on demand? Or do they have only a fixed selection? Are technical assistants available to teach and help people with seed collection, seed beds, nurseries, site preparation or planting? Are training materials available? Are there any demonstration or training facilities?

Enough seed sources and nurseries may exist to supply the community. However, most groups will have to collect some of their own seeds or at least grow some of their own seedlings. Find out from the group about the places where it might be possible to grow seedlings—on private, public or group land with a reliable water source.

If possible, arrange for someone from the community to accompany you to nearby seed centres or nurseries to see the number and quality of plants available. Before you leave the group, invite people to return to you with additional information or ideas which may occur to them later. Also, if they know of specialists who are not present, invite them to bring these people to you or to take you to them. If the group is willing, ask them to meet again for a short time after 1 or 2 weeks to review the list of species and uses and to add any new information or opinions after discussing these topics with their friends, families and community groups.

APPENDIX IV: SUMMARIZING INFORMATION FOR PLANNING AGROFORESTRY RESEARCH AND EXTENSION

A summary of information for planning agroforestry research and extension should be prepared, adapted and shared with the community. This makes it possible to review, discuss and revise the results of the surveys with those who have contributed and who will act upon the results.

Summarizing information may be more difficult than collecting it. However, there are ways to keep summaries simple and clear. Begin by making a general summary for the whole community. Review all your notes and interview summaries, such as those in Tables 2 and 3 of Appendix I and Parts 1, 2 and 3 of Appendix III. Fill out the form below or your own summary sheet. Distribute this to the community in whatever way is locally appropriate and meet with representative groups to discuss and revise the results. Distribute the revised summaries and keep them as project records.

Repeat this exercise for each sub-group of the community which you have identified as having distinct interests and potential for agroforestry activities. For each community you will then have one general summary and a number of separate summaries for each **type** of group (not for each group interviewed).

1. Background Information

List the most important categories of land type, the most important land uses, the major land-user groups and the most serious problems at the household and community level.

1.1. Land types

____________________ ____________________
____________________ ____________________
____________________ ____________________
____________________ ____________________

1.2. Major land uses

____________________ ____________________
____________________ ____________________
____________________ ____________________
____________________ ____________________

1.3. Major land-user groups

____________________ ____________________
____________________ ____________________
____________________ ____________________
____________________ ____________________
____________________ ____________________

1.4. Most important resource/production problems at the household level (in order of importance)

Resource	*Production*
______________	______________
______________	______________
______________	______________

1.5. Most important resource/production problems at the community level (in order of importance)

Resource	*Production*
______________	______________
______________	______________
______________	______________
______________	______________

2. Information Related to Tree Planting/Planning

Based on prior interview notes and observations, list the most likely species and agroforestry practices compatible with local conditions, priorities and resources.

2.1. Highest-priority uses (products/services) of trees, shrubs and grasses (in order of importance)

____

____

____

2.2. Most popular (preferred or important) species for each use

Use	*Species*
________	______________
	______________
	______________
	______________
	______________
________	______________
	______________
	______________
	______________
	______________
________	______________
	______________
	______________
	______________

2.3. Obstacles to tree planting or agroforestry practices in the community

Obstacle	*Changeable? yes/no/how?*

2.4. Places or types of place where serious confusion, conflict or overlapping rights to land and trees exists and groups involved
At the household level

Nature of conflict	*Where*	*Between whom and whom*

At the community level

Nature of Conflict	*Where*	*Between whom and whom*

2.5. Most likely locations for future tree planting and agroforestry practices (in order of preference)

2.6. Most likely and practical agroforestry practices—given the planting places, uses and species listed above—and groups most likely to practice

Practice	*Who*

3. Work Plan

After everyone has had time to think and talk among themselves, meet with the groups again to choose the species and agroforestry practices they would like to try. Select the places, species and practices for research or extension programmes, as appropriate, and begin a work plan for the next season.

Place/space	*Agroforestry practice*	*Species*

4. Sample of Group Interview Summary

Site: Ginga, Lake Victoria shore, Kenya.
Ecozone: 4.
Group: Ginga Women's Group.

Profile of group: Low-income farmers of fishing village want to supplement cash income; formal group is recent, previously on-farm labour exchange.

Number present: 19 people from 13 households.

Profile of those present: leader, technical assistant, cross-section of members, 5 men (husbands of members, interested elders).

Household types: few women-headed households; women farm and men fish, but men help with farm and women process and market fish; extended families and polygamous family units; shared compounds, each adult woman manages own plot with husband.

Farming system types: subsistence mixed cropping with some cash crops; sell cotton and maize; sell some fruits; one crop per year; most have cattle, goats, sheep; among all present, 4 oxen and 1 donkey; most cultivation by hand, some use/rent oxen.

Land tenure: most have 0.5 hectare or less, about one-third have 0.5 to 2 hectares; about one-third have own grazing land; heavy dependence on road ways, border lands and land belonging to others; grazing is free, fuelwood collection restricted; land adjudicated; 100-metre strip on lake shore is public (water and land for tree nursery).

Crops grown (C = cash sale; S = subsistence): sorghum S (preferred); maize S (poor yields last 4 years); beans S; cowpeas S; green gram S; groundnuts CS; cotton C; citrus S; banana S; mango S; tamarind CS; papaya S; *Balanites* S; fig S; all buy some grain March-June; about half sell some food crops at harvest; stored food is women's property, crops in field are men's property.

Cropping calendar and distribution of labour (M = men, W = women, C = children):

Jan	Feb	Mar	Apr	May	Jun	Jul	Aug	Sep	Oct	Nov	Dec
very dry		prepare land and plant		weed		harvest		dry		very dry	
		M, W		W		W, C					

Fuelwood trees/shrubs, ranked by preference: *Balanites aegyptiaca* 1; *Acacia seyal* 2; *Tamarindus indica*; *Ficus* spp.; *Euphorbia tirucalli* (prefer to replace with less aggressive fencerow plant, less smoky wood); *Lantana camera* (widely used, not disliked).

Timber trees: *Markhamia platycalyx, Cassia siamea* (for poles; some sell); *Albizia coriaria* (for boats, purchased from Uganda); purchase *Olea africana* from lake islands; going prices are KShs 7 for roof poles and KShs 10 for wall poles.

Fodder trees/shrubs/grasses: *Grewia trichocarpa* (cattle, sheep and goats); *Balanites aegyptiaca* (cattle, sheep and goats); *Markhamia platycalyx* (cattle, sheep and goats in dry season); *Cassia siamea* (goats in very dry periods); *Euphorbia tirucalli* (goats in very dry periods); *Tamarindus indica* (sheep, goats and cattle in dry periods); *Lantana camera* (cattle, sheep and goats in dry season)

Plants used for commercial or domestic handicrafts: *Balanites aegyptiaca* (spoons); *Markhamia platycalyx* (furniture, preferred); *Albizia coriaria* (furniture, preferred); *Cassia siamea, Eucalyptus* spp. (furniture); purchase palm fibre for baskets from Uganda; use *Balanites* and *Grewia* to build storehouses; *Indigofera spicata* for bird or hen cages, baskets, trays; *Tinnea arthipica* for trays.

Medicinal plants and wild foods: *Harrisonia abyssinica; Tinnea aethiopica*?

Trees planted on cropland: *Markhamia platycalyx, Tamarindus indica* left on cropland; papaya planted on cropland; *Cassia siamea* occurs, but they say it has negative effect; other are neutral or improve crops.

Problems, ranked from most to least serious: 1 rainfall, 2 fishing catch (fish population disturbed, Nile perch and tilapia dominate), 3 cash, 4 fuelwood, 5 cattle fodder, 6 land shortage (contributes to 3, 4 and 5).

Receptive (all, most, few, none) to tree/shrub planting on: cropland (few), boundaries (all), woodlots (few), grazing land (few), home compound (most).

Species suggested: *Cassia, Eucalyptus* (to grow on boundaries, in lots for fuel, poles); *Markhamia platycalyx, Tamarindus indica* (in cropland); *Balanites aegyptiaca; Ficus* spp. (to grow on cropland, in fences, home compounds for fuelwood, fruit, fodder).

Potential farm-trial sites: lots of interest; most willing to try living fences; group leader's home available for demonstration; group members will provide 2 or 3 more farm sites.

Plans, objectives for nursery: vegetables for home use and sale; seedlings for home, sale; want fruit, firewood, poles.

Comments, other information: heavy use of fuel for smoking fish; pay KShs 5-8 per headload for firewood, KShs 3 per headload for sisal poles; use one headload of fuelwood to smoke 20 fish; interest in replacing *Euphorbia tirucalli* in hedgerows with fast-growing, more manageable fuelwood species; men are present, employed and actively interested; economy of farm/fishing families is fairly commercial; interest high in selling tree products or substituting own tree products for those now purchased, especially fuelwood for smoking fish.

Note: check commercial fuelwood sources in terms of **future** *economy, impact, alternative enterprises.*

APPENDIX V: GLOSSARY OF TERMS

acid (soil): A soil having a pH of less than 7.0. A soil with a higher pH is called basic or alkaline. Some plants, such as *Leucaena leucocephala*, do not grow well in strongly acid soils (pH less than 5.5). Limestone (a basic rock) is often ground and added to soils to reduce their acidity. (See alkaline.)

agroforestry: The deliberate use of woody perennials (trees, shrubs, palms, bamboos) on the same land-management unit as agricultural crops, pastures and/or animals. This may consist of a mixed spatial arrangement in the same place at the same time, or a sequence over time.

agropastoralism: The combination of herbaceous crops and extensive livestock production in the same land-use system.

agrosilviculture: The combination of herbaceous crops and trees or shrubs in the same land-use system.

agrosilvopastoralism: The combination of trees or shrubs, herbaceous crops and livestock production in the same land-use system.

alkaline (soil): Any soil with a pH greater than 7.0. Also called basic. (See acid.)

alley cropping or alley farming: The practice of growing annual crops in the spaces between rows of trees or hedgerows. This is sometimes called hedgerow intercropping.

annual: A plant that grows for only one season (or year) before dying, in contrast to a perennial, which grows for more than one season.

arid: A climate characterized by so little rainfall that cultivation is only possible if supported by water management. For the purposes of this book, an arid area has an average annual rainfall of less than 200 mm. Koeppen codes BW and BW2.

basic (soil): See alkaline.

berm: See bund.

biomass: The weight of material produced by a living organism or collection of organisms. The term biomass is usually applied to plants. It may include the entire plant, or it may be qualified to include only certain parts of the plant, e.g. aboveground or leafy biomass. Biomass is expressed in terms of fresh weight or dry weight.

black cotton soil: Vertisol. A dark soil with a high clay content most common in low-lying flat areas. This soil type is sticky and gummy when wet and develops large cracks when dry. It is usually alkaline to neutral and the clay content is high in montmorillonite. Special techniques are often needed to grow crops or trees in this type of soil.

boma (Kiswahili): Fence of living or dead branches used to enclose livestock or to direct their movements or to protect human settlements from wild animals. Also called zeriba (West Africa) or corral (USA).

browse: The buds, shoots, leaves and flowers of woody plants which are eaten by livestock or wild animals.

budding: The practice of splicing a bud from one tree into the bark of another, usually to obtain high-quality fruit on hardy, established trees.

bund: A ridge of earth placed in a line along the contour of a slope to control water runoff and soil erosion.

bush: 1. A small woody plant (see shrub); 2. Uncleared, wild landscape with scattered vegetation.

cambium: Cellular tissue of plants from which growth occurs.

compost: Organic residues from plants and animals, sometimes mixed with soil, that are piled, moistened and allowed to decompose. Used to improve soil fertility when incorporated into the soil, often in small, intensive gardens. See also mulch, used in a related practice.

clump: A close grouping of stems of trees, bushes or grasses.

contour: Line joining all places at the same height above sea-level. Used on maps to indicate change in elevation, or the slope of land.

controlled grazing: Livestock grazing limited to specific areas, often relying in part on fodder which is cut and brought to the animals.

coppicing: Cutting certain tree species close to ground level to produce new shoots from the stump. Also occurs naturally in some species if the trees are damaged.

corral: See boma.

cross-section: A way of drawing an object, scene or landscape by cutting a slice through it to illustrate its various parts and their relation to each other.

crown: The canopy or top of a single tree or other woody plant that carries its main branches and leaves at the top of a fairly clean stem.

cutting: A piece of a branch or root cut from a living plant with the objective of developing roots and growing a new plant, genetically identical to the original parent (a clone).

cut-and-carry: Fodder or other plant products which are harvested and carried to a different location to be used or consumed.

deciduous: A plant which loses all or a part of its leaves at the end of a season's growth. The opposite of evergreen.

delineate: To trace the outline of an area.

demarcate: To mark a line or boundary.

dialogue: A discussion in which there is clear communication between two individuals or groups on the basis of equality and respect.

direct seeding: Sowing seeds directly where they are to develop into mature plants.

dispersed: Individuals (such as plants, animals or people) scattered across an area; spread widely, rather than concentrated in one or a few locations.

ditch: Long, narrow trench or channel.

ecosystem: All the plants and animals in a given area and their physical environment, including the interactions between them.

erosion: The wearing away of the land surface by running water, wind, ice or movement due to gravity.

evergreen: Plants which retain their leaves and remain green throughout the year. Opposite of deciduous.

extensive: Land use or management spread over a large area where land is plentiful (at least for those who control it). Opposite of intensive.

exotic: A plant or animal species which has been introduced outside its natural range. Opposite of indigenous.

fallow: Land resting from cropping, which may be grazed or left unused, often colonized by natural vegetation.

family: A taxonomic category between order and genus. Plants or animals in the same family share some common characteristics.

farming system: All the elements of a farm which interact as a system, including people, crops, livestock, other vegetation, wildlife, the environment and the social, economic and ecological interactions between them.

floodplain: Land flooded periodically by a river, stream, lake or pond.

fodder: Parts of plants which are eaten by domestic animals. These may include leaves, stems, fruit, pods, bark, flowers, pollen or nectar.

foliage: The mass of leaves of trees or bushes.

grafting: The practice of propagating trees by taking a small shoot from one tree and attaching it to another so that the cambium layers from both are in contact and the transferred shoot grows as part of the main tree. This is normally used to obtain high-quality fruit from hardy, well established trees (rootstock).

graze: To feed on grass.

green manure: Green leafy material applied to the soil to improve its fertility.

groundcover: Living or non-living material which covers the soil surface.

groundwater: Water which is underground. It may be pumped to the surface or reached by plant roots or wells or may feed into bodies of surface water.

gully: A deep, narrow channel cut into the soil by erosion.

gully erosion: The removal of soil by water concentrated in deep, narrow channels.

genus: A taxonomic category between family and species. A genus consists of one or more closely related species and is defined largely in terms of the characteristics of the flower and/or fruit.

hedgerow: (or hedge) A closely planted line of shrubs or small trees, often forming a boundary or fence.

herbaceous: A plant that is not woody and does not persist above ground beyond one season.

herbivore: An animal that feeds only on plants.

home garden: A complex collection of woody and herbaceous plants deliberately grown in small plots in or near home compounds, often associated with the production of small domestic animals.

indigenous: Native to a specific area; not introduced. Opposite of exotic.

infiltration: The movement of water into the soil.

infiltration ditch: A trench or ditch, sometimes filled with stony or loosely packed material, that traps water and allows it to seep into the soil.

intensive: Land use or management concentrated in a small area of land. Opposite of extensive.

intercept: To catch rainwater or surface runoff water with either vegetation or structures. On the surface, this prevents further runoff and erosion downslope and generally increases the amount of water available for use by plants, animals or rural households.

intercropping: Growing two or more crops in the same field at the same time in a mixture.

interface: The area where there is positive or negative interaction between two entities, such as between a row of trees and a row of crops.

interspersed: Where different entities, such as plant types, are mixed together rather than kept in separate, distinct areas.

landscape: An area of land, usually between 10 and 100 square kilometres, including vegetation, built structures and natural features, seen from a particular viewpoint. Landscape ecologists and landscape designers use this term differently from the more popular definition used in this text.

land-use system: The way in which land is used by a particular group of people within a specified area.

latex: The milky sap of some plant species. It has many traditional and industrial uses and is often toxic.

litter: The uppermost layer of organic material on the soil surface, including leaves, twigs and flowers, freshly fallen or slightly decomposed.

lop: To cut one or more branches of a standing tree.

microcatchment: A small earthwork structure designed to catch and direct rainfall to a growing plant, crop, household or livestock watering place. Used in arid and semi-arid areas to encourage plant growth. See also rainwater harvesting.

microclimate: The temperature, sunlight, humidity and other climatic conditions in a small localized area, for example in one field, stand of trees or in the vicinity of a given plant.

monoculture: A community of plants all of a single species, generally artificially established.

mulch: Plant or non-living materials used to cover the soil surface with the object of protecting the soil from the impact of rainfall, controlling weeds or moisture loss and, in some cases, fertilizing the soil. See also compost, used in a related practice.

multipurpose tree (MPT): A woody perennial which is grown to provide more than one product or service.

multistoreyed: Relating to a vertical arrangement of plants so that they form distinct layers, from the herbaceous layer to the uppermost tree canopy.

niche: A place in the landscape which is suited to a particular plant, animal or activity because of the local social and/or ecological environment.

nitrogen-fixing: Relating to a plant that has the ability to convert nitrogen in the air into a form which can be used by plants. This process is performed by another organism that lives within the roots of the plant. In leguminous plants the organism is a bacterium. In other plants, such as *Casuarina* species, it is an actinomycete.

overstorey: The highest layer of vegetation, often the tree canopy, which grows over lower shrub or plant layers.

palatable: Desirable to eat.

perennial: A plant that grows for more than one year, in contrast to an annual, which grows for only one year (or season) before dying.

permeable: Allowing the movement of air, water or other material. In soils, refers to conditions favourable to the movement of water into and through the soil.

pole: The stem of a young tree before its crown begins to expand, or large shoots resulting from coppicing or pollarding. In Africa, cut poles are often used for house construction.

pollarding: Cutting back the crown of a tree in order to harvest wood and browse to produce regrowth beyond the reach of animals and/or to reduce the shade cast by the crown.

propagation: Reproduction of plants by seed, cuttings, roots or other means.

pruning: Cutting back plant growth, including side branches or roots.

rainwater harvesting: Use of small earthwork structures or bunds to trap runoff water from rainfall and concentrate it in a small area to increase the water available for plant growth. When water from streams or other sources is also being captured, this method may be called water harvesting. This same term is sometimes used (not in this text) to refer to tanks and rooftop catchment structures for collecting rainwater. See also microcatchment.

recharge: Rainfall or water in rivers, streams, ponds or lakes that seeps down through the soil and replenishes the groundwater.

regeneration: Regrowth.

ridge: A long raised strip of earth.

rill erosion: The removal of soil by water in many small channels a few inches deep.

root sucker: A shoot arising from the root of a plant.

rotation: In agriculture, changing the crops grown on a particular piece of ground from season to season. In forestry, the length of time between establishment and harvesting of a plantation or tree.

runoff: Rainfall or other water that does not infiltrate into the soil but flows across the surface.

saline soil: A soil that contains enough salts to interfere with the growth of most crop plants. The pH is usually less than 8.5, exchangeable sodium less than 15% and conductivity of the saturation extract over 4 mS/cm (formerly mmho/cm).

sapling: A young tree, no longer a seedling but not yet a pole, a few metres high at most and about 2.5 cm in diameter at breast height.

savannah: Extensive grass-covered plain, with some scattered trees or shrubs; **dry savannah**: semi-arid shrub plain with 250 to 600 mm average annual rainfall; **moist savannah**: subhumid shrub or tree plain with 600 to 1500 mm average annual rainfall.

seed treatment: Nicking, soaking in water or treating seeds with substances such as insecticides or fungicides to improve germination.

selective cutting: Cutting specific, undesirable types of plants and leaving the rest to grow.

semi-arid: In this book, semi-arid refers to a climate with average annual rainfall of 150 to 900 mm. In semi-arid areas, rainfall in some years is insufficient to maintain crop cultivation. See also steppe climate.

semi-permeable: Allowing some limited movement of air, water or other substance. In soils, refers to condition that limit the movement of water into and through the soil.

service: Shade, soil improvement or other benefit, other than a removable product, derived from trees.

sheet erosion: The removal of a fairly uniform layer of soil from the land surface by runoff water.

shrub: A woody plant that remains less that 10 metres tall and produces shoots or stems from its base (see bush).

shoot: A stem; may also refer to new growth of a plant, usually including a stem.

silvopastoralism: The combination of trees or shrubs with pasture and/or livestock production in the same land-use system.

slope: The inclination or angle of the land surface, which can be measured as a percent, ratio or in degrees or grades.

social fencing: Cultural or social regulation that protects an area of land.

soil moisture: Water in the soil, a portion of which is available to plants.

species: A taxonomic category below genus. Individuals within a species can interbreed, but breeding between species does not normally occur or results in sterile offspring (hybrids).

staggered (planting, harvesting): Referring to activities carried out at different times or locations, instead of synchronized to occur at the same time or place.

steppe climate: A treeless plain with short, sparse grass, generally with rainfall averaging less than 250 mm a year (see also semi-arid).

subhumid climate: In the tropics, a climate with rainfall averaging 900 to 1200 mm a year and susceptible to drought. Also known as 'grassland' climate.

superficial roots: Plant roots near the soil surface.

tap root: A persistent, and often enlarged, main plant root that grows downward.

tannin: A substance often extracted from tree bark, among other sources, and used to tan animal hides.

tenure: The right to property, granted by custom and/or law, which may include land, trees and other plants, animals and water.

termite resistant: Not likely to be attacked by termites or unlikely to suffer damage if attacked.

tree: A woody plant with one main trunk and a more-or-less distinct and elevated head.

trench: See ditch.

tied ridge: A narrow strip of ground left unexcavated between contour furrows or trenches, which joins—at intervals—parallel ridges on the contour. Tied ridges are meant to contain runoff water from the ridges on the contour.

till: Cultivate the soil; open and turn the soil surface.

timber: Wood of a sufficient size to be used for heavy construction and/or cut into boards.

topography: The physical description of land; changes in elevation due to hills, valleys and other features.

understorey: The lower layer of vegetation, often grasses, shrubs or crops, that grows under other vegetation.

watershed: A unit of the landscape that contains all the drainage areas and channels contributing to a single stream or river system.

wilding (or wildling) A young plant growing naturally without having been planted or seeded by people.

woody: Plants which consist in part of wood; not herbaceous.

zero-grazing Livestock production systems in which the animals are fed in pens or other confined areas and are not permitted to graze.

APPENDIX VI: CONTACTS

GOVERNMENT RESEARCH AND DEVELOPMENT AGENCIES AND EDUCATIONAL INSTITUTIONS

Direction des Eaux et Forêts
B.P. 7044
Ouagadougou, Burkina Faso

Institut des Sciences Agronomiques du Burundi (ISABU)
(M. Oscar Ndabikingiye, Director General)
B.P. 795
Bujumbura, Burundi

Forestry Research Centre
P.O. Box 1034
Addis Ababa, Ethiopia

Federal Research Centre for Forestry and Forest Products
(Prof. H.J. von Maydell)
Leuschnerstr. 91
2050 Hamburg 80,
Federal Republic of Germany

Central Soil and Water Conservation Research and Training Institute
Dehra Dun, U.P. 248195, India

Forest Research Institute
P.O. New Forest
Dehra Dun, U.P., India

Indian Council of Agricultural Research (ICAR)
Krishi Bhavan
Dr. Rajandra Prasad Rd.
New Delhi 110001, India

Kenya Agricultural Research Institute (KARI)
P.O. Box 74
Kikuyu, Kenya

Kenya Forestry Research Institute (KEFRI)
Dryland Agroforestry Project
P.O. Box 340
Machakos, Kenya

Kenya Forestry Research Institute (KEFRI)
(Dr. J. Odera, Director)
P.O. Box 20412
Nairobi, Kenya

Kenya Prisons
Horticulturist/Nurseries
P.O. Box 30175
Nairobi, Kenya

Ministry of Agriculture and Livestock Development
Department of Agriculture
P.O. Box 30028
Nairobi, Kenya

Ministry of Energy and Regional Development
P.O. Box 30582
Nairobi, Kenya

Tree Seed Programnme
Ministry of Energy and Regional Development
P.O. Box 21552
Nairobi, Kenya

Department of Soil Science
University of Nairobi
P.O. Box 30197
Nairobi, Kenya

College of Agriculture
and Veterinary Sciences
University of Nairobi
Kabete Campus
P.O. Box 30197
Nairobi, Kenya

Department of Agricultural Engineering
University of Nairobi
P.O. Box 30197
Nairobi, Kenya

Agroforestry Commodity Team Research
(Mr. Ommar Ali Itimu)
P.O. Box 158
Lilongwe, Malawi

Forestry Research Institute of Malawi
P.O. Box 270
Zomba, Malawi

Direction des Eaux et Forêts
B.P. 275
Bamako, Mali

Ministère de l'Agriculture
(Directeur du Projet Productivité)
Niamey, Niger

Direction des Eaux et Forêts
B.P. 578
Niamey, Niger

Centre d'Information pour le
Développement Rural
Ministère de l'Agriculture
Niamey, Niger

Forestry Commission
(Dr. Jonathan Okafor)
P.M.B. 1028
Enugu, Anambra State, Nigeria

Institut des Sciences Agronomiques
du Rwanda (ISAR)
(Mr. Léopold Gahamanyi, Director)
B.P. 138
Butare, Rwanda

Direction Générale des Forêts
B.P. 621
Kigali, Rwanda

Institut Sénégalais de
Recherches Agricoles
Centre National de Recherches, Parc
Forestier de Hann
B.P. 2312
Dakar, Senegal

Direction des Eaux et Forêts
B.P. 1831
Dakar, Senegal

National Range Agency
P.O. Box 1759
Mogadishu, Somalia

Tanzania Forestry Research Institute
P.O. Box 95
Lushoto, Tanzania

Department of Forest Biology
Sokoine University of Agriculture
(Dr. Luther Lulandala)
P.O. Box 3010
Morogoro, Tanzania

Department of Forest Economics
and Policy
Sokoine University of Agriculture
(Dr. Aku Oktingati)
P.O. Box 3010
Morogoro, Tanzania

Forest and Beekeeping Division
Ministry of Natural Resources
and Tourism
(Mr. E.M. Mnzava)
P.O. Box 426
Dar es Salaam, Tanzania

Faculty of Agriculture and Forestry
Makerere University
(Professor J.S. Mugerwa, Dean)
P.O. Box 7062
Kampala, Uganda

Institute of Development Studies
(Dr. Robert Chambers)
University of Sussex
Brighton BN1 9RE, U.K.

Oxford Forest Institute
(Dr. Peter Wood,
Dr. Jeffrey Burley, Director)
South Parks Road
Oxford OX1 3RB, U.K.

Social Forestry Network
Agricultural Administrative Unit
Overseas Development Institute (ODI)
Regent's College
Inner Circle, Regent's Park
London NW1 4NS, U.K.

Department of Forestry and
Wood Science
University College of North Wales
(Prof. L. Roche)
Bangor, Gwynedd LL 57 2UW, U.K.

Caesar Kleberg Wildlife
Research Institute
Texas A. & M University
(Dr. Peter Felker)
Campus Box 218
Kingsville, Texas 78363, U.S.A.

Land Tenure Center
University of Wisconsin
1300 University Avenue
Madison, Wisconsin 53706, U.S.A.

Office of Technology Assessment
U.S. Congress
600 Pennysylvania Ave., S.E.
Washington, D.C. 20510, U.S.A.

International Forestry Staff
United States Department of
Agriculture/Forest Service (USDA/FS)
Room 1208 RPE
P.O. Box 2419
Washington, D.C. 20013, U.S.A.

Forestry Support Program
United States Forest Service
FSP Room 1208 RPE
P.O. Box 2417
Washington, D.C. 20013, U.S.A.

Forest Research Division
P.O. Box 22099
Kitwe, Zambia

Soil Productivity Research Programme
Ministry of Agriculture
Mt. Makulu Station
Private Bag 7
Chilanga, Zambia

Agroforestry Specialists
Technical and Extension Services
Department of Agriculture
(Mr. Newton Spicer)
P.O. Box 8117
Causeway, Harare, Zimbabwe

Forestry Commission
(Dr. Yemi Katerere,
Deputy Commissioner for
Rural Afforestation)
P.O. Box 8111
Causeway, Harare, Zimbabwe

BILATERAL AID AGENCIES

Danish International Development
Agency (DANIDA)
Mutomo Project
P.O. Box 41
Mutomo, Kitui, Kenya

Forestry Advisor
Forestry Department
Embassy of Switzerland
P.O. Box 30752
Nairobi, Kenya

Norwegian Agency for International Development (NORAD)
(Mr. G.C. Barrow)
Norwegian Embassy
P.O. Box 46363
Nairobi, Kenya

Overseas Development Administration (ODA)
Embu-Meru-Isiolo Project (EMI)
P.O. Box 1199
Embu, Kenya

Regional Office
Soil Conservation Programme
Swedish International Development Authority (SIDA)
P.O. Box 30600
Nairobi, Kenya

OTAPS/Forestry and Natural Resources
Peace Corps
806 Connecticut Ave., N.W.
Washington, D.C. 20526, U.S.A.

Bureau for Africa
U.S. Agency for International Development (USAID)
Washington, D.C. 20523, U.S.A.

Office of Science and Technology
U.S. Agency for International Development (USAID)
S&T/FENR Rm 515-D SA-IB
Washington, D.C. 20523, U.S.A.

INTERNATIONAL RESEARCH AND DEVELOPMENT ORGANIZATIONS

Benin FSR Programme
Semi-Arid Food Grain Research and Development of the OAU (SAFGRAD)
(Dr. Robert Mike Otsyina)
B.P. 3
N'Dali, Benin

Projects and Programs Division
Comité Permanent Inter-Etats de Lutte contre la Sécheresse dans le Sahel (CILSS)
B.P. 7049
Ouagadougou, Burkina Faso

Burkina FSR Programme
Semi-Arid Food Grain Research and Development of the OAU (SAFGRAD)
(M. Amadou Ibra Niang)
B.P. 1783
Ouagadougou, Burkina Faso

International Coordinator and Field Officer
Semi-Arid Food Grain Research and Development of the OAU (SAFGRAD)
B.P. 1783
Ouagadougou, Burkina Faso

Chef du Projet d'Agroforesterie
(Dr. Bahiru Duguma)
IRA-ICRAF
B.P. 2067
IRA Nkolbisson
Yaounde, Cameroon

Cameroon FSR Programme
Semi-Arid Food Grain Research and Development of the OAU (SAFGRAD)
(Dr. Wangoi Migongo-Bake)
B.P. 415
Garoua, Cameroon

International Livestock Centre for Africa (ILCA)
P.O. Box 5689
Addis Ababa, Ethiopia

Centre Technique Forestier Tropical (CTFT)
(M. Jean Pierre Goudet)
45 bis, Avenue de la Belle Gabrielle
94736 Nogent-sur-Marne Cedex, France

Club du Sahel
39, Boulevard Suchet
75775 Paris Cedex 16, France

Centre Louis Emberger (CNRS)
(Dr. H.N. Le Houérou)
Route de Mende
B.P. 5051
34060 Montpellier Cedex, France

Institut Français de Recherche Scientifique pour le Développement en Coopération (ORSTOM)
213, rue La Fayette
75480 Paris Cedex 10, France

Policy and Planning Service
Forestry Department
(Ms. Marilyn Hoskins,
Community Forestry Officer)
Food and Agriculture Organization of the United Nations (FAO)
Via delle Termi di Caracalla
00100 Rome, Italy

Energy/Development International (EDI)
P.O. Box 39002
Nairobi, Kenya

Environment Liaison Centre (ELC)
P.O. Box 72461
Nairobi, Kenya

International Council for Research in Agroforestry (ICRAF)
P.O. Box 30677
Nairobi, Kenya

Pasture Network for Eastern and Southern Africa (PANESA)
ILCA-Kenya
P.O. Box 46847
Nairobi, Kenya

Agroforestry Research Network for Africa (AFRENA)
International Council for Research in Agroforestry (ICRAF)
(Dr. J. Maghembe)
Makoka Agricultural Research Station
Private Bag 3
Thondwe, Malawi

Sahelian Center
International Crops Research Institute for the Semi-Arid Tropics (ICRISAT)
(Dr. Rick J. Van den Belt)
B.P. 12404
Niamey, Niger

Centre Technique Forestier Tropical (CTFT)
B.P. 12090
Niamey, Niger

International Institute of Tropical Agriculture (IITA)
(Dr. B.T. Kang, Dr. B.N. Okigbo)
P.M.B. 5320
Ibadan, Nigeria

Non-Governmental Liaison Service
International Tree Project Clearinghouse (ITPC)
2 U.N. Plaza
DC-2-RM 1103
New York, NY 10017, U.S.A

PRIVATE VOLUNTARY AND NON-GOVERNMENTAL ORGANIZATIONS

Baobab Farm, Ltd.
P.O. Box 90202
Mombasa, Kenya

Catholic Diocese of Nakuru
P.O. Box 938
Nakuru, Kenya

Catholic Diocese of Nakuru
East Pokot Agricultural Project, Kositei
P.O. Marigat
via Nakuru, Kenya

Regional Technical Advisor
for Agroforestry
CARE International in Kenya
P.O. Box 43864
Nairobi, Kenya

Green Belt Movement
(Dr. Wangari Mathai,
Programme Coordinator)
National Council of Women of Kenya
P.O. Box 43741
Nairobi, Kenya

Kenya Energy Non-Government
Organizations (KENGO)
(Mr. Achoka Aworry, Director)
P.O. Box 48197
Nairobi, Kenya

Kenya Woodfuel Development
Programme (KWDP)
P.O. Box 56121
Nairobi, Kenya

Mazingira Institute
P.O. Box 14550
Nairobi, Kenya

Environment and Development Action
in the Third World (ENDA)
B.P. 3370
Dakar, Senegal

Concern
P.O. Box 704
Iringa, Tanzania

Tanga Integrated Rural Development
Programme
P.O. Box 72
Lushoto, Tanzania

Joint Energy and Environment Projects
(JEEP)
P.O. Box 1684
Jinja, Uganda

Royal Botanic Gardens
Kew, Wakehurst Place, Ardingly
Haywards Health
West Sussex RH17 67N, U.K.

Agriculture and Natural Resources
Program
CARE International
660 First Avenue
New York, N.Y. 10015, U.S.A.

International Institute for Environment
and Development (IIED)
Suite 302
1717 Massachusetts Ave., N.W.
Washington, D.C. 20004, U.S.A.

International Institute for Environment
and Development (IIED)
3 Endsleigh Street
London WC1H ODD, U.K.

International Tree Crops Institute
P.O. Box 888
Winters, California 95694, U.S.A.

New York Botanical Gardens
Bronx, N.Y. 10458, U.S.A.

Nitrogen Fixing Tree Association
(Prof. James J. Brewbaker)
P.O. Box 680
Waimanolo, Hawaii 97795, U.S.A.

Agroforestry Program
Winrock International Institute
for Agricultural Development
RT 3, Petit Jean Mountain
Morriton, Arkansas 72110, U.S.A.

Environment and Development
Activities (ENDA)-Zimbabwe
P.O. Box 3492
Harare, Zimbabwe

Oxfam Regional Headquarters
P.O. Box 4590
Harare, Zimbabwe

APPENDIX VII: LIST OF ACRONYMS

CARE: Cooperative for American Relief Everywhere (New York, NY, USA)

CTFT: Centre Technique Forestière Tropicale (Nogent-sur-Marne, France)

ENDA: Environnement et Développement du Tiers Monde (Dakar, Senegal)

FAO: The Food and Agricultural Organization of the United Nations (Rome, Italy)

GTZ: Gesellschaft für Technische Zusammenarbeit (Eschborn, Federal Republic of Germany)

ICAR: Indian Council of Agricultural Research (Central Arid Zone Research Institute, Jodphur, India)

ICRAF: International Council for Research in Agroforestry (Nairobi, Kenya)

IIED: International Institute for Environment and Development (London, U.K.)

IITA: International Institute for Tropical Agriculture (Ibadan, Nigeria)

ILCA: International Livestock Centre for Africa (Addis Ababa, Ethiopia)

ILO: International Labour Organization of the United Nations (Geneva, Switzerland)

IPAL/UNESCO: Integrated Project for Arid Lands of the United Nations Educational, Scientific and Cultural Organization (Marsabit, Kenya)

ISRA: Institut Sénégalais de Recherche Agricole (Dakar, Senegal)

KARI: Kenya Agricultural Research Institute (Nairobi, Kenya)

KEFRI: Kenya Forestry Research Institute (Muguga, Kenya)

KENGO: Kenya Energy and Environment Organizations (Nairobi, Kenya)

KREDP: Kenya Renewable Energy Development Project (Nairobi, Kenya)

NAS: National Academy of Sciences (Washington, D.C., U.S.A.)

NORAD: Norwegian Agency for International Development (Oslo, Norway)

SIDA: Swedish International Development Authority (Stockholm, Sweden)

UNDP: United Nations Development Programme (New York, N.Y., U.S.A.)

UNESCO: United Nations Educational, Scientific and Cultural Organization (Paris, France)

UNICEF: United Nations Children's Fund (New York, N.Y., U.S.A.)

USAID: United States Agency for International Development (Washington, D.C., U.S.A.)

USDA: United States Department of Agriculture (Beltsville, Maryland, U.S.A.)

VITA: Volunteers for International Technical Assistance (Arlington, Virginia, U.S.A.)

WMO: World Meteorological Organization (Geneva, Switzerland)

APPENDIX VIII: REFERENCE LIST

Ahmed, A. (1979). Effects of land use on soil characteristics in the Sudan. In H.O. Mongi and P.A. Huxley, eds. *Soil research in agroforestry: proceedings of an expert consultation*. Nairobi: ICRAF, pp. 1-13.

Ahmed, F.A. and Ahmed, A.E. (1978). Plant cover, nutritive value and carrying capacity of a range pasture in the low-rainfall woodland savannah of Western Sudan. In *East African Agricultural and Forestry Journal*. 43: (3): 238-45.

All Africa Press Service (1986). *Tanzania: legume replacing inorganic fertilizers*. Nairobi: All Africa Press Service.

Amare Getahun and Bashir Jama (1986). *Alley farming in the coastal area of Kenya*. Nairobi: Ministry of Energy, Kenya Renewable Energy Development Project, 26 pp.

Anger, G., ed. (1985). *L'arbre et la haie dans l'exploitation paysanne*. Fiche Technique No. 3. Nyabisindu, Rwanda: Project Agropastoral de Nyabisindu, 126 pp.

Anger, G., Duvel, K.H., Ntwali, J.M.V. and Pietrowicz, P. (1985). Recherche et vulgarisation dans le domaine agroforestier au projet agropastoral de Nyabisindu. *Bulletin Agricole du Rwanda*. April, 89-93.

Arap Sang, F.K., Hoekstra, D.A. and Okumu, R. (1985). *Preliminary results of on-station green manure trials with leaves from* Leucaena leucocephala, Cassia siamea *and* Terminalia brownii. Research Report No. 1. Kikuyu, Kenya: KEFRI, Dryland Agroforestry Research Project, 21 pp.

Atta-Krah, A.N. and Francis, P.A. (1987). The role of on-farm trials in the evaluation of composite technologies: the case of alley farming in southern Nigeria. *Agricultural Systems*. 23: 133-52.

Bala Subramanian, V. and Agli, A. (1986). The role of agroforestry in the farming systems in Rwanda with special reference to the Bugeresa-Gisaka-Migongo (BGM) region. *Agroforestry Systems*. 4: 271-89.

Barrow, E.G.C. (1984). *NORAD's involvement in the forestry sector of Turkana District: past work and future prospects*. Lodwar, Kenya: Forest Department, 23 pp.

Barrow, E.G.C. (1985). *ASAL/TRDP quarterly report for the forestry sector, Turkana District*. Lodwar, Kenya: Forestry Department, 23 pp.

Barrow, E.G.C., ed. (1987). *Forestry manual for primary teachers in Turkana District*. Lodwar, Kenya: Forestry Department, 51 pp.

Barrow, E.G.C. (forthcoming). Value of traditional knowledge in present-day soil conservation practice: the example of the Pokot and Turkana. In D.J. Thomas, ed. *Proceedings of the Third National Soil Conservation Workshop*. Nairobi: University of Nairobi, Agricultural Engineering Department.

Baumer, M. (1983). *Notes on trees and shrubs in arid and semi-arid regions*. Rome: FAO, 270 pp.

Baumer, M. (1987). *Le role possible de l'agroforesterie dans la lutte contre la desertification et la degradation*. Brussels: European Economic Community, 240 pp.

Bognettean-Verlinden, E. 1980. *Study on impact of windbreaks in Majjia Valley, Niger.* Wageningen, The Netherlands: Agricultural University. 77 pp.

Boudet, G.C. and Toutain, B. (1980). The integration of browse plants within pastoral and agropastoral systems in Africa. In H.N. Le Houerou, ed. *Browse in Africa: the current stage of knowledge*. Addis Ababa: ILCA, pp. 427-32.

Buck, L., ed. (1981). *Proceedings of the Kenya National Seminar on Agroforestry*. Nairobi: ICRAF and the University of Nairobi, 638 pp.

Buck, L. (1988). *Agroforestry extension training sourcebook*. New York: Cooperative for American Relief Everywhere (CARE) International, 540 pp.

Buck, L. and Ngugi, A., eds. (1988). *Agroforestry monitoring and evaluation project: regional workshop report*. New York: CARE International, 235 pp.

Chambers, R. (1981). Rapid rural appraisal: rationale and repertoire. *Public Administration and Development*. 1:95-106.

Chambers, R. (1983). *Rural development:: putting the last first.*. Harlow, UK: Longman, 246 pp.

Chambers, R. and Jiggins, J. (1987a). Agricultural research for resource-poor farmers. Part I: transfer-of-technology and farming systems research. *Agricultural Administration and Extension*. 27:35-52.

Chambers, R. and Jiggins, J. (1987b). Agricultural research for resource-poor farmers. Part II: a parsimonious paradigm. *Agricultural Administration and Extension*. 27: 109-28.

Charreau, C. and Vidal, P. (1965). Influence de l'*Acacia albida* Del. sur le sol, nutrition minérale et rendements des mils *pennisetum* au Sénégal. *Agronomie Tropicale*. 20:601-26.

Cheruiyot, H.K., Broberg, G., Wamae, L.W. and Wachira, T.M. (1981). Effects of eucalyptus leaves on the survival of aquatic snails. *East African Agricultural and Forestry Journal.* 46 (4): 77-80.

Clanet, J.C. and Gillet, H. (1980). *Commiphora africana*, browse tree of the Sahel. In H.N. Le Houerou, ed. *Browse in Africa: the current stage of knowledge.* Addis Ababa: ILCA, pp. 443-45.

Conway, G. and McKracken, J. (1988). *Agroecosystems analysis.* London: IIED, 35 pp.

Dale, I.R. and Greenway, P.J. (1961). *Kenya trees and shrubs.* Nairobi: Buchanan's Kenya Estate Ltd., 654 pp.

Darnhofer, T. (1983). Microclimatic effects and design considerations of shelterbelts. In D.A. Hoekstra, ed. *Agroforestry systems for small-scale farmers: proceedings of an ICRAF/BAT workshop.* Nairobi: ICRAF, pp. 95-111.

Dennison, S. (1986). *Majjia Valley windbreak evaluation: Niger briefing document.* Niamey: CARE, 7 pp.

Depommier, D. (1985). The role of the woody vegetation in soil conservation and rehabilitation. In *Proceedings of a workshop on soil conservation on grazing lands.* Nairobi: Ministry of Agriculture and Livestock Development, Soil and Water Conservation Division, pp. 82-118.

Douglas, M.G. (1984). *Integrated land use.* Lilongwe: Ministry of Agriculture, 16 pp.

Douglas, M.G. (1988). *Integrating conservation into the farming system: land-use planning for smallholder farmers—concepts and procedures.* London: Commonwealth Secretariat, 130 pp.

Dressler, J. and Neumann, I. (1982). Agriculture de couverture du sol. *Bulletin Agricole du Rwanda.* Kigali, Rwanda.

ENDA (no date). *Environnement africain: plantes et arbes utiles.* Fiche Technique Numbers 207-68. Dakar: ENDA.

Farrington, J., ed. (1988). *Farmer participatory research.* Reprinted from *Experimental Agriculture*, 24: 269-342. London: Overseas Development Institute, 73 pp.

Feldstein, H., Buck, L. and Rocheleau, D. (in press). The Siaya agroforestry project in Kenya. In H. Feldstein, ed. *Intra-household dynamics and farming systems research and extension: a casebook.* Hartford, Connecticut, USA: Kumarian Press, 165 pp.

Felker, P. (1981). Uses of tree legumes in semi-arid areas. *Economic Botany*. 35 (2): 174-86.

Fernandes, E.C.M., Oktingati, A. and Maghembe, J. (1984). The Chagga home gardens: a multistoried agroforestry cropping system on Mt. Kilimanjaro (Northern Tanzania). *AgroforestrySystems*. 2: 73-86.

Feuerstein, M. (1986). *Partners in evaluation: evaluating development and community programmes with participants*. London: MacMillan, 240 pp.

Finkel, M., ed. (1985). *Turkana water-harvesting manual.* Nairobi: prepared by Finkel and Finkel Ltd. for NORAD, 135 pp.

Fliervoet, E. (1982). *An inventory of trees and shrubs in the Northern Division of Machakos District, Kenya*. Wageningen, The Netherlands: Wageningen Agricultural University and Nairobi: ICRAF, 32 pp.

FAO (1977). *Guidelines for watershed management.* FAO Conservation Guide No 1. Rome: FAO, 293 pp.

FAO (1978a). *China: forestry support for agriculture.* Forestry Paper No. 12. Rome: FAO, 103 pp.

FAO (1978b). *Forestry for local community development.* Forestry Paper No. 7. Rome: FAO, 114 pp.

FAO (1986a). *Brise-vent et rideaux abris avec référence particulière aux zones sèches.* Cashier FAO Conservation No. 15. Rome: FAO, 385 pp.

FAO (1986b). *Tree growing by rural people.* Forestry Paper No. 64. Rome: FAO, 130 pp.

Fortmann, L. and Bruce, J., eds. (1988). *Whose trees?* Boulder, Colorado: Westview Press, 341 pp.

Fortmann, L. and Riddell, J. (1985). *Trees and tenure: an annotated bibliography for agroforesters and others.* Nairobi: ICRAF and Madison, Wisconsin, USA: Land Tenure Center, 135 pp.

Fortmann, L. and Rocheleau, D. (1985). Women and agroforestry: four myths and three case studies. *Agroforestry Systems*. 2: 253-72.

France, Ministère de Coopération (1979). *Conservation des sols au sud du Sahara.* Paris: CTFT, 296 pp.

Gupta, R. (1979). *Plants for environmental conservation.* Dehra Dun, India: Bishen Singh Mahendra Pal Singh, 247 pp.

Hedfors, L. (1981). *Evaluation and economic appraisal of soil conservation in a pilot area.* Nairobi: Ministry of Agriculture, 24 pp.

Henderson, L. (1983). Barrier plants in South Africa. *Bothalia.* 14 (3-4): 635-39.

Hudson, N. (1976). *Soil Conservation.* London: Batsford, 320 pp.

Huxley, P.A., Rocheleau, D.E. and Wood, P.A. (1985). *Farm systems and agroforestry research in northern Zambia.* Nairobi: ICRAF, 76 pp.

ICRAF (1983a). *Guidelines for agroforestry diagnosis and design.* Working Paper No. 6. Nairobi: ICRAF, 25 pp.

ICRAF (1983b). *Resources for agroforestry diagnosis and design.* Working Paper No. 7. Nairobi: ICRAF, 383 pp.

ICRAF (1987). Agroforestry for the Control of Soil Erosion. *ICRAF Newsletter.* 19: 1-2.

IITA (no date). *Alley cropping: a stable alternative to shifting cultivation.* Ibadan, Nigeria: IITA, 25 pp.

ILCA (1986). *PanesaNewsletter.* Addis Ababa: ILCA, Pasture Network for Eastern and Southern Africa.

Jensens, J.W. (1986). *A landscape development plan for Kakayuni, Kenya: landscape planning and design with the application of agroforestry in the rehabilitation of eroded grazing lands.* M.Sc. thesis. Wageningen, The Netherlands: Department of Landscape Architecture, Agricultural University and Machakos, Kenya: Dryland Agroforestry Research Project, 165 pp.

Jiggins, J. and Feldstein, H. (in press). *Methodologies casebook on intra-household issues in farming systems research.* Hartford, Connecticut, USA: Kumarian Press, 180 pp.

Kang, B.T. and Duguma, B. (1984). *Nitrogen management in alley cropping systems.* Ibadan, Nigeria: IITA, 18 pp.

Kuchar, P. (1981). *The plants of Kenya: a handbook of uses and ecological status.* Nairobi: Kenya Rangeland Ecological Monitoring Unit (KREMU), 397 pp.

Lamprey, H.F. (1983). UNESCO's Integrated Project on Arid Lands (IPAL): a silvopastoral approach. In P. Huxley, ed. *Plant research and agroforestry.* Nairobi: ICRAF, 129-36.

Lamprey, H.F., Herlocker, D.J. and Field, C.R. (1980). Report on the state of knowledge on browse in East Africa. In H. Le Houerou, ed. *Browse in Africa: the current state of knowledge*. Addis Ababa: ILCA, pp. 33-54.

Le Houerou, H. (1980a). Agroforestry techniques for the conservation and improvement of soil fertility in arid and semi-arid zones. In H. Le Houerou, ed. *Browse in Africa: the current state of knowledge*. Addis Ababa: ILCA, pp. 433-35.

Le Houerou, H., ed. (1980b). *Browse in Africa: the current state of knowledge*. Addis Ababa: ILCA, 491 pp.

Leblond, B. and Guerin, L. (1983). *Soil conservation: project design and implementation using labour-intensive techniques*. Geneva: ILO, 206 pp.

Mbonye, A. (no date). *Sesbania sesban*. Kenya Indigenous Tree Training Series. Nairobi: Kenya Energy and Environment Organizations, 21pp.

Miehe, S. (1986). *Acacia albida* and other multipurpose trees on the Fur farmlands in the Jebel Marra Highlands, Western Darfur, Sudan. *Agroforestry Systems*. 4: 89-119.

Mongi, H.O. and Huxley, P.A., eds. (1979). *Soils research in agroforestry: proceedings of an expert consultation*. Nairobi: ICRAF, 585 pp.

Nair, P.K.R. (1985). Classification of agroforestry systems. *Agroforestry Systems*. 3: 97-128.

Nair, P.K.R. (1987). *Agroforestry systems in major ecological zones of the tropics and subtropics*. Working Paper No. 47. Nairobi, ICRAF, 15 pp.

NAS (1979). *Tropical legumes: resources for the future*. Washington, DC: National Academy Press, 331 pp.

NAS (1980). *Firewood crops: shrub and tree species for energy production*. Volumes 1 and 2. Washington, DC: National Academy Press, 329 pp.

NAS (1983). Calliandra: *a versatile small tree for the humid tropics*.. Innovations in Tropical Reforestation Series. Washington, DC: National Academy Press, 52 pp.

NAS (1984). *Casuarinas: nitrogen-fixing trees for adverse sites*. Innovations in Tropical Reforestation Series. Washington, DC: National Academy Press, 118 pp.

Okafar, J.C. (1980). Edible indigenous woody plants in the rural economy of the Nigerian forest zone. *Forest Ecology Management*. 3: 45-55.

Onyewotu, L.O.Z. and Owonubi, J.J. (1987). Effect of structural variation in shelterbelts on windflow: a case study of a Sudan savannah environment in Nigeria. In *Windbreak and shelterbelt technology for increasing agricultural production.* Washington, DC: USAID, pp. 87-89.

Oresdain, L.E. and Barrow, E.G.C. (forthcoming). Soil conservation along Lodwar-Kapuna road project. In D.J. Thomas, ed. *Proceedings of the Third National Soil Conservation Workshop.* Nairobi: University of Nairobi, Agricultural Engineering Department.

Parkash, R. and Hocking, D. (1986). *Some favourite trees for fuel and fodder.* New Delhi: Society for Promotion of Wastelands Development, 187 pp.

Parry, M.S. (1953). Tree planting in Tanganyika. 2. Species for highlands. *East African Agricultural and Forestry Journal.* 19: 89-102.

Parry, M.S. (1954). Tree planting in Tanganyika. 3. Species for dry areas. *East African Agricultural and Forestry Journal.* 19: 154-60.

Parry, M.S. (1953). Tree planting in Tanganyika. 4. Species for coastal areas. *East African Agricultural and Forestry Journal.* 19: 49-53.

Pereira, H. (1983). *Land use and water resources.* London: Cambridge University Press, 246 pp.

Poschen, P. (1986). An evaluation of the *Acacia albida*-based agroforestry practices in the Hararghe highlands of Eastern Ethiopia. *Agroforestry Systems.* 4: 129-43.

Poulsen, G. (1981). *Malawi: the function of trees in small-farmer production systems.* Rome: FAO, 65 pp.

Poulsen, G. (1985). Trees on cropland: preserving an African heritage. *CERES (FAO Review on Agriculture and Development).* 18 (2): 24-28.

Poynton, R.J. (1985). *Characteristics and uses of selected trees and shrubs cultivated in South Africa.* Pretoria: South African Forestry Research Institute, 202 pp.

Raintree, J., ed. (1987a). *D&D user's manual: an introduction to agroforestry diagnosis and design.* Nairobi: ICRAF, 110 pp.

Raintree, J., ed. (1987b). *Land, trees and tenure: proceedings of an international workshop on tenure issues in agroforestry.* Nairobi: ICRAF and Madison, Wisconsin: Land Tenure Center, 412 pp.

Redhead, J.F. and Maghembe, J.A. (1980). *Agroforestry: preliminary results of intercropping* Acacia, Eucalyptus *and* Leucaena *with maize and beans*. Morogoro, Tanzania: Forestry Division, University of Dar Es Salaam, 19 pp.

Redhead, J.F. and Maghembe, J.A. (no date). *Agroforestry: integrated land use for rural communities in tropical Africa*. Morogoro, Tanzania: Forestry Division, University of Dar Es Salaam, 19 pp.

Richards, P. (1985). *Indigenous agricultural revolution: ecology and food production in West Africa*. London: Hutchinson, 192 pp.

Rocheleau, D.E. (1985a). *Land-use planning with rural farm households and communities: participatory agroforestry research*. Working Paper No. 36. Nairobi: ICRAF, 43 pp.

Rocheleau, D.E. (1985b). *Criteria for re-appraisal and re-design: intra-household and between-household aspects of FSRE in three Kenyan agroforestry projects*. Working Paper No. 37. Nairobi: ICRAF, 24 pp.

Rocheleau, D.E. (1987). The user perspective and the agroforestry research and action agenda. In H.L. Gholz, ed. *Agroforestry: realities, possibilities and potentials*. Dordrecht, The Netherlands: Martinus Nijhoff and ICRAF, pp. 59-87.

Rocheleau, D.E. (1988). *A community-based approach to agroforestry research and development*. Nairobi: Ford Foundation, 120 pp.

Rocheleau, D.E. and Malaret, L. (1988). Use of ethnoecology in agroforestry systems research: an example of agroforestry technology and pest management research in Kenya. In W. Miller, ed. *Proceedings of the annual farming systems research symposium*. Morrillton, Arkansas, USA: Winrock International.

Rocheleau, D.E. and van den Hoek, A. (1984). *The application of ecosystems and landscape analysis in agroforestry diagnosis and design: a case study from Kathama Sublocation, Machakos District, Kenya*. Working Paper No. 11. Nairobi: ICRAF, 92 pp.

Rocheleau, D.E., Wachira, K., Malaret, L. and Wanjohi, B.M. (in press). Local knowledge, innovations and ethnoecological methods to complement traditional knowledge and farmer innovation in agroforestry. In A. Pacey, R. Chambers and L. Thrupp, eds. *Farmer innovation and agricultural research*. Brighton, UK: University of Sussex, Institute of Development Studies, 280 pp.

Sands, E.B., Thomas, D.B., Knight, J. and Pratt, D.J. (1970). Preliminary selection of pasture plants for the semi-arid areas of Kenya. *East African Agricultural and Forestry Journal*. 35: 49-57.

Scherr, S. and Mueller, E. (1988). *Technology monitoring and evaluation in agroforestry projects: a review*. Nairobi: ICRAF, 40 pp.

Scoones, I. (1988). *Community management of indigenous woodland project: Zvishavane and Chivi Districts, Zimbabwe*. Harare: ENDA, 115 pp.

Seignobos, C. (1980). Des fortifications végétales dans la zone Soudano-sahélienne (Tchad et Nord-Cameroun). *Cahiers ORSTOM*. Série Sciences Humaines. 17, 3-4:191-222.

Shankar, V. (1984). Inter-relationships of tree overstorey and understorey vegetation in silvi-pastoral system. In K.A. Shankarnarayan, ed. *Agroforestry in arid and semi-arid zones*. Jodphur, India: ICAR, Central Arid Zone Research Institute, pp. 143-49.

Shankarnarayan, K.A., ed. (1984). *Agroforestry in arid and semi-arid zones*. Jodphur, India: ICAR, Central Arid Zone Research Institute, 295 pp.

Sheng, T.C. (1979). Protection of cultivated slopes: terracing steep slopes in humid regions. In *Guidelines for watershed management*. FAO Conservation Guide 1. Rome: FAO, pp. 147-79.

Sokoine University of Agriculture (1986). *Research report on Marajea* (Crotalaria ochroleuca). Morogoro, Tanzania: Sokoine University of Agriculture.

Ssekabembe, C.K. (1983). *Studies on soil moisture depletion from* Leucaena leucocephala-Cassia siamea *hedgerows in dryland areas*. Nairobi: ICRAF, 14 pp.

Stollen, K.A. (1983). *Peasants and agricultural change in northern Zambia*. Occasional Paper No. 4. As, Norway: Department of Agricultural Economics, Agricultural University of Norway.

Storrs, A.E.G. (1979). *Know your trees: some of the common trees found in Zambia*. Ndola, Zambia: Forest Department, 380 pp.

Storrs, A.E.G. (1982). *More about trees: interesting facts and uses of some common Zambian trees*. Ndola, Zambia: Forest Department, 127 pp.

Storrs, A.E.G. and Pierce, G.D. (1982). *Don't eat these: a guide to some local poisonous plants and mushrooms.*. Ndola, Zambia: Forest Department, 53 pp.

Sumberg, J. and Okali, C. (1988). Farmers, on-farm research and the development of new technology. *Experimental Agriculture*. 24:333-42.

Synott, T.J. (1979). *A report on prospects, problems and proposals for tree planting*. Nairobi: IPAL/UNESCO.

Tanzania, Ministry of Lands, Natural Resources and Tourism (1984). *Trees for village forestry.* Dar es Salaam: Ministry of Lands, Natural Resources and Tourism, Forest Division, 125 pp.

Teel, W. (1984). *A pocket directory of trees and shrubs in Kenya.* Nairobi: KENGO. 151 pp.

Tejwani, K.G. (1979). Soil fertility status maintenance and conservation for agroforestry systems on wasted land in India. In H.O. Mongi and P.A. Huxley, eds. (1979). *Soils research in agroforestry: proceedings of an expert consultation.* Nairobi: ICRAF, pp. 141-74.

USAID (1987). *Windbreak and shelterbelt technology for increasing agricultural production.* Washington, DC: USAID, 219 pp.

USDA (1954). *A manual on conservation of soil and water.* Agricultural Handbook No. 61. Washington, DC: USDA, 350 pp.

Valet, S. (1985). *Action villageois bénévole de lutte anti-érosive dans les unités expérimentales de Thysse-Sonkorong en 1984.* Sine Saloum, Senegal: ISRA, 63 pp.

Vietmeyer, N. (1985). In praise of shrubs. *CERES (FAO Review on Agriculture and Development).* 18 (2): 28-32.

von Carlowitz, P. (1986). *Multipurpose tree and shrub seed directory.* Nairobi: ICRAF, 265 pp.

von Maydell, H.J. (1983). *Arbes et arbustes du Sahel: leurs caractéristiques et leurs utilisations.* Eschborn, FR Germany: GTZ, 530 pp.

Wachiira, K.K. (1987). *Women's use of off-farm and boundary lands: agroforestry potentials.* Nairobi: ICRAF, 137 pp.

Weber, F. with Storey, C. (1986). *Reforestation in arid lands.* Arlington, Virginia, USA: VITA, 335 pp.

Weber, F. and Hoskins, M. (1983a). *Agroforestry in the Sahel.* Blacksburg, Virginia, USA: Virginia Polytechnic Institute, 102 pp.

Weber, F. and Hoskins, M. (1983b). *Soil conservation technical sheets.* Moscow, Idaho, USA: University of Idaho, 124 pp.

Wenner, C.G. (1980). *An outline of soil conservation in Kenya.* Nairobi: Ministry of Agriculture, Soil Conservation and Extension Unit, 57 pp.

Wenner, C.G. (1981). *Soil conservation in Kenya: especially in small-scale farming in high-potential areas using labour-intensive methods.* Seventh revised edition. Nairobi: Ministry of Agriculture and SIDA, 230 pp.

Wijewardene, R. and Waidyanatha, P. (1984). *Conservation farming: systems, techniques and tools for small farmers in the humid tropics.* Colombo, Sri Lanka: Department of Agriculture, 38 pp.

Wilson, G. and Kang, B.T. (1981). Developing stable and productive biological cropping systems for the humid tropics. In B. Stonehouse, ed. *Biological husbandry: a scientific approach to organic farming.* London: Butterworth, 193-203.

Wilson, K. (1987). *Community management of indigenous woodland project: report to ENDA, Zimbabwe.* Harare: ENDA, 115 pp.

WMO (1964). *Windbreaks and shelterbelts.* Technical Note No. 59. Geneva: WMO Secretariat, 188 pp.

Yamoah, C.F., Ay, P. and Agboola, A.A. (1986). The use of *Gliricidia sepium* for alley cropping in the southern guinea savanna zone of Nigeria. *International Tree Crops Journal.* 3:267-79.

Young, A. (1986). *The potential of agroforestry for soil conservation. Part I: erosion control.* Working Paper No. 42. Nairobi: ICRAF, 68 pp.

Young, A. (1987). *The potential of agroforestry for soil conservation. Part II: maintenance of fertility.* Working Paper No. 43. Nairobi: ICRAF, 135 pp.

INDEX

A

B

D

G

H

I

J

K

R

S

T

U

V

W

Y

Z